Natural Disaster Research, Prediction and Mitigation Series

EARTHQUAKES: RISK, MONITORING AND RESEARCH

NATURAL DISASTER RESEARCH, PREDICTION AND MITIGATION SERIES

The Phoenix of Natural Disasters: Community Resilience
Kathryn Gow and Douglas Paton (Editors)
2008. ISBN: 978-1-60456-161-6

Natural Disasters: Public Policy Options for Changing the Federal Role in Natural Catastrophe Insurance
GAO
2008. ISBN: 978-1-60456-717-5

Solar Activity and Forest Fires
Milan Rodovanovic and Joao Fernando Pereira Gomes
2009. ISBN: 978-1-60741-002-7

National Emergency Responses
Paul B. Merganthal (Editor)
2009. ISBN: 978-1-60692-354-2

Earthquakes: Risk, Monitoring and Research
Earl V. Leary (Editor)
2009. ISBN: 978-1-60692-648-2

Natural Disaster Research, Prediction and Mitigation Series

EARTHQUAKES: RISK, MONITORING AND RESEARCH

EARL V. LEARY
EDITOR

Nova Science Publishers, Inc.
New York

Copyright © 2009 by Nova Science Publishers, Inc.

All rights reserved. No part of this book may be reproduced, stored in a retrieval system or transmitted in any form or by any means: electronic, electrostatic, magnetic, tape, mechanical photocopying, recording or otherwise without the written permission of the Publisher.

For permission to use material from this book please contact us:
Telephone 631-231-7269; Fax 631-231-8175
Web Site: http://www.novapublishers.com

NOTICE TO THE READER

The Publisher has taken reasonable care in the preparation of this book, but makes no expressed or implied warranty of any kind and assumes no responsibility for any errors or omissions. No liability is assumed for incidental or consequential damages in connection with or arising out of information contained in this book. The Publisher shall not be liable for any special, consequential, or exemplary damages resulting, in whole or in part, from the readers' use of, or reliance upon, this material.

Independent verification should be sought for any data, advice or recommendations contained in this book. In addition, no responsibility is assumed by the publisher for any injury and/or damage to persons or property arising from any methods, products, instructions, ideas or otherwise contained in this publication.

This publication is designed to provide accurate and authoritative information with regard to the subject matter covered herein. It is sold with the clear understanding that the Publisher is not engaged in rendering legal or any other professional services. If legal or any other expert assistance is required, the services of a competent person should be sought. FROM A DECLARATION OF PARTICIPANTS JOINTLY ADOPTED BY A COMMITTEE OF THE AMERICAN BAR ASSOCIATION AND A COMMITTEE OF PUBLISHERS.

LIBRARY OF CONGRESS CATALOGING-IN-PUBLICATION DATA

ISBN: 978-1-60692-648-2

Published by Nova Science Publishers, Inc. ✤ *New York*

CONTENTS

Preface		vii
Chapter 1	Earthquakes: Risk, Monitoring, Notification, and Research *Peter Folger*	1
Chapter 2	Annual Report of the National Earthquake Hazards Reduction Program *FEMA, National Institute of Standards and Technology, National Science Foundation, and USGS*	29
Chapter 3	Forecasting California's Earthquakes—What Can We Expect in the Next 30 Years?	113
Index		119

PREFACE

Close to 75 million people in 39 states face some risk from earthquakes. Seismic hazards are greatest in the western United States, particularly California, Alaska, Washington, Oregon, and Hawaii. The Rocky Mountain region, a portion of the central United States known as the New Madrid Seismic Zone, and portions of the eastern seaboard, particularly South Carolina, also have a relatively high earthquake hazard. Compared to the loss of life in other countries, relatively few Americans have died as a result of earthquakes over the past 100 years. The United States, however, faces the possibility of large economic losses from earthquake damaged buildings and infrastructure.

Chapter 1 - Close to 75 million people in 39 states face some risk from earthquakes. Seismic hazards are greatest in the western United States, particularly California, Alaska, Washington, Oregon, and Hawaii. The Rocky Mountain region, a portion of the central United States known as the New Madrid Seismic Zone, and portions of the eastern seaboard, particularly South Carolina, also have a relatively high earthquake hazard. Compared to the loss of life in other countries, relatively few Americans have died as a result of earthquakes over the past 100 years. The United States, however, faces the possibility of large economic losses from earthquake- damaged buildings and infrastructure.

Until Hurricane Katrina in 2005, the 1994 Northridge (CA) earthquake was the costliest natural catastrophe to strike the United States; some damage estimates were $26 billion (in 2005 dollars). Estimates of total loss from a hypothetical earthquake of magnitude more than 7.0 reach as high as $500 billion for the Los Angeles area. The May 12, 2008, magnitude 7.9 earthquake in Sichuan, China, has raised some concerns about the possibility of a similar devastating earthquake occurring in seismically active and densely populated parts of the United States, such as California.

Given the potentially huge costs associated with a severe earthquake, an ongoing issue for Congress is whether the federally supported programs aimed at reducing U.S. vulnerability to earthquakes are an adequate response to the earthquake hazard. Under the National Earthquake Hazards Reduction Program (NEHRP), four federal agencies have responsibility for long-term earthquake risk reduction: the U.S. Geological Survey (USGS), the National Science Foundation (NSF), the Federal Emergency Management Agency (FEMA), and the National Institute of Standards and Technology (NIST). They variously assess U.S. earthquake hazards, send notifications of seismic events, develop measures to reduce earthquake hazards, and conduct research to help reduce overall U.S. vulnerability to earthquakes.

Congress established NEHRP in 1977, and its early focus was on research that would lead to an improved understanding of why earthquakes occur and to an ability to predict their occurrence precisely. Understanding has improved about why and where earthquakes occur; however, reliably predicting the precise date and time an earthquake will occur is not yet possible. Congress most recently reauthorized NEHRP in 2004 (P.L. 108-360) and authorized appropriations through FY2009. The 2004 reauthorization designated NIST as the lead agency to create better synergy among the agencies and improve the program. Congress may wish to determine whether the reorganized structure has yielded expected benefits for the program. Appropriations for NEHRP have not met levels authorized for the past four years, falling short by an average of 31% for FY2005 through FY2008. What effect funding at the levels enacted through FY2008 has had on the U.S. capability to detect earthquakes and minimize losses after an earthquake occurs is not clear.

Chapter 2 - Of all natural hazards threatening the United States, earthquakes pose the greatest risk in terms of casualties and damage. According to a 2006 National Research Council (NRC) report, 42 states have regions with some degree of earthquake potential and 18 states have areas of high or very high seismicity. Over 75 million people live in urban areas with moderate to high earthquake risk. The NRC report also notes that the estimated value of structures in all states prone to earthquake damage is approximately $8.6 trillion (2003 dollars).

Chapter 3 - In a new comprehensive study, scientists have determined that the chance of having one or more magnitude 6.7 or larger earthquakes in the California area over the next 30 years is greater than 99%. Such quakes can be deadly, as shown by the 1989 magnitude 6.9 Loma Prieta and the 1994 magnitude 6.7 Northridge earthquakes. The likelihood of at least one even more powerful quake of magnitude 7.5 or greater in the next 30 years is 46%—such a quake is most likely to occur in the southern half of the State. Building codes, earthquake insurance, and emergency planning will be affected by these new results, which highlight the urgency to prepare now for the powerful quakes that are inevitable in California's future.

In: Earthquakes: Risk, Monitoring and Research
Editor: Earl V. Leary

ISBN: 978-1-60692-648-2
© 2009 Nova Science Publishers, Inc.

Chapter 1

EARTHQUAKES: RISK, MONITORING, NOTIFICATION, AND RESEARCH[*]

Peter Folger

ABSTRACT

Close to 75 million people in 39 states face some risk from earthquakes. Seismic hazards are greatest in the western United States, particularly California, Alaska, Washington, Oregon, and Hawaii. The Rocky Mountain region, a portion of the central United States known as the New Madrid Seismic Zone, and portions of the eastern seaboard, particularly South Carolina, also have a relatively high earthquake hazard. Compared to the loss of life in other countries, relatively few Americans have died as a result of earthquakes over the past 100 years. The United States, however, faces the possibility of large economic losses from earthquake- damaged buildings and infrastructure.

Until Hurricane Katrina in 2005, the 1994 Northridge (CA) earthquake was the costliest natural catastrophe to strike the United States; some damage estimates were $26 billion (in 2005 dollars). Estimates of total loss from a hypothetical earthquake of magnitude more than 7.0 reach as high as $500 billion for the Los Angeles area. The May 12, 2008, magnitude 7.9 earthquake in Sichuan, China, has raised some concerns about the possibility of a similar devastating earthquake occurring in seismically active and densely populated parts of the United States, such as California.

---

[*] Excerpted from CRS Report RL33861, dated June 19, 2008.

Given the potentially huge costs associated with a severe earthquake, an ongoing issue for Congress is whether the federally supported programs aimed at reducing U.S. vulnerability to earthquakes are an adequate response to the earthquake hazard. Under the National Earthquake Hazards Reduction Program (NEHRP), four federal agencies have responsibility for long-term earthquake risk reduction: the U.S. Geological Survey (USGS), the National Science Foundation (NSF), the Federal Emergency Management Agency (FEMA), and the National Institute of Standards and Technology (NIST). They variously assess U.S. earthquake hazards, send notifications of seismic events, develop measures to reduce earthquake hazards, and conduct research to help reduce overall U.S. vulnerability to earthquakes.

Congress established NEHRP in 1977, and its early focus was on research that would lead to an improved understanding of why earthquakes occur and to an ability to predict their occurrence precisely. Understanding has improved about why and where earthquakes occur; however, reliably predicting the precise date and time an earthquake will occur is not yet possible. Congress most recently reauthorized NEHRP in 2004 (P.L. 108-360) and authorized appropriations through FY2009. The 2004 reauthorization designated NIST as the lead agency to create better synergy among the agencies and improve the program. Congress may wish to determine whether the reorganized structure has yielded expected benefits for the program. Appropriations for NEHRP have not met levels authorized for the past four years, falling short by an average of 31% for FY2005 through FY2008. What effect funding at the levels enacted through FY2008 has had on the U.S. capability to detect earthquakes and minimize losses after an earthquake occurs is not clear.

The 1994 Northridge (CA) earthquake caused as much as $26 billion (in 2005 dollars) in damage, according to one estimate, and was one of the costliest natural disasters to strike the United States. The Federal Emergency Management Agency (FEMA) has estimated that earthquakes cost the United States over $4 billion per year. Some cost estimates of a single, large earthquake striking the Los Angeles area range as high as $500 billion. A hypothetical scenario for a magnitude 7.8 earthquake in southern California, released on May 22, 2008, estimated a possibility of 1,800 fatalities and over $200 billion in economic losses. The May 12, 2008, magnitude 7.9 earthquake in Sichuan, China, has thus far resulted in nearly 70,000 fatalities. (Sources for these data are provided below.)

Under the National Earthquake Hazards Reduction Program (NEHRP), the federal government supports efforts to assess and monitor earthquake hazards and risk in the United States. Four federal agencies, responsible for long-term earthquake risk reduction, coordinate their activities under NEHRP: the U.S.

Geological Survey (USGS), the National Science Foundation (NSF), FEMA, and the National Institute of Standards and Technology (NIST). Congress reauthorized NEHRP in 2004 (P.L. 108-360), and authorized appropriations for a total of $902.5 million over five years.

Given the potentially huge costs associated with a large, damaging earthquake in the United States, an ongoing issue for Congress is whether the federally supported earthquake programs are adequate for the earthquake risk. This report describes estimates of earthquake hazards and risk in the United States, the current federal programs that support earthquake monitoring and that provide notification after a seismic event, and the programs that support mitigation and research aimed at reducing U.S. vulnerability to earthquakes.

EARTHQUAKE HAZARDS AND RISK

Figure 1 indicates that good information exists on where earthquakes are likely to occur and how severe the earthquake magnitude and resulting ground shaking are likely to be. The map in Figure 1 depicts the potential shaking hazard from future earthquakes. It is based on the frequency at which earthquakes occur in different areas and how far the strong shaking extends from the source of the earthquake. In Figure 1, the hazard levels indicate the potential ground motion — expressed as a percentage of the acceleration due to gravity (g) — with up to a 1 in 10 chance of being exceeded over a 50-year period.

All 50 states are vulnerable to earthquake hazards, although risks vary greatly across the country. Seismic hazards are greatest in the western continental United States, particularly California, Washington, Oregon, and Alaska and Hawaii. Alaska is the most earthquake-prone state, experiencing a magnitude 7 earthquake[1] almost every year and a magnitude 8 earthquake every 14 years on average. Because of its low population and infrastructure density, Alaska has a relatively low risk for large economic losses from an earthquake. In contrast, California has more citizens and infrastructure at risk than any other state because of the state's frequent seismic activity combined with its high population.

Figure 1 also shows relatively high earthquake hazard in the Rocky Mountain region, portions of the eastern seaboard — particularly South Carolina — and a part of the central United States known as the New Madrid Seismic Zone (discussed below). Other portions of the eastern and northeastern United States are also vulnerable to moderate seismic hazard. According to the USGS, 75 million people in 39 states are subject to significant risk. During the period 1975-1995, only four states did not experience detectable earthquakes: Florida, Iowa,

North Dakota, and Wisconsin. (The map shown in figure 1 is based on the new USGS seismic hazards map, released on April 21, 2008. See box for more information about the new map.)

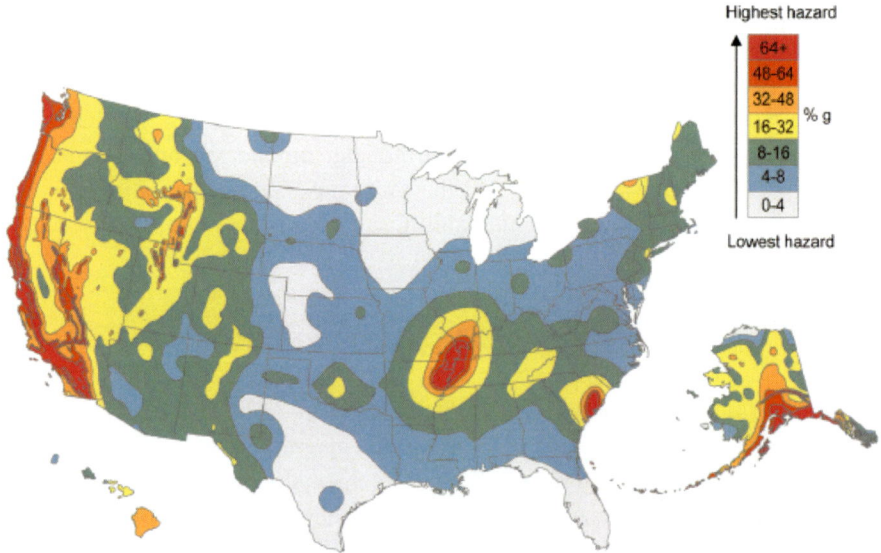

Source: USGS Fact Sheet 2008-3018 (April 2008), at [http://earthquake.usgs.gov/research/hazmaps/products_data/images/nshm_us02.gif]. Modified by CRS.

Figure 1. Earthquake Hazard in the United States.

Shaking hazards maps, such as the one in Figure 1, are often combined with other data, such as the strength of existing buildings, to estimate possible damage in an area following an earthquake. The combination of seismic risk, population, and vulnerable infrastructure can help improve the understanding of which urban areas across the United States face risks from earthquake hazards that may not be immediately obvious from the probability maps of shaking hazards alone. The USGS has identified 26 urban areas that face a significant seismic risk from the combination of population and severity of shaking. Table 1 lists those areas at greatest risk.

USGS RELEASES NEW NATIONAL SEISMIC HAZARDS MAPS AND NEW EARTHQUAKE FORECAST FOR CALIFORNIA

On April 21, 2008, the USGS released new National Seismic Hazards Maps that update the version published in 2002. Compared to the 2002 version, the new maps indicate lower ground motions (by 10% to 25%) for the central and eastern United States, based on modifications to the ground-motion models used for earthquakes. The new maps indicate that estimates of ground motion for the western United States are as much as 30% lower for certain types of ground motion, called long-period seismic waves, which affect taller, multistory buildings. Ground motion that affects shorter buildings of a few stories, called short-period seismic waves, is roughly similar to the 2002 maps. The new maps show higher estimates for ground motion for western Oregon and Washington compared to the 2002 maps, due to new ground motion models for the offshore Cascadia subduction zone. In formulating the 2008 maps, the USGS gave more weight to the probability of a catastrophic magnitude 9 earthquake occurring along the Cascadia subduction zone. That fault ruptures, on average, every 500 years, and has the potential to generate destructive earthquakes and tsunamis along the coasts of Washington, Oregon, and northern California.

According to a report released on April 14, 2008, California has a 99% chance of experiencing a magnitude 6.7 or larger earthquake in the next 30 years. The likelihood of an even larger earthquake, magnitude 7.5 or greater, is 46% and will likely occur in the southern part of the state. The fault with the highest probability of generating at least one earthquake of magnitude 6.7 or greater over the next 30 years is the San Andreas in southern California (59% probability); for northern California it is the Hayward-Rodgers Creek Fault (31%). The earthquake forecasts are not predictions (i.e., they do not give a specific date or time), but represent probabilities over a given time period, and even the probabilities have variability associated with them. The earthquake forecasts are known as the "Uniform California Earthquake Rupture Forecast (UCERF)" and are produced by a working group comprised of the USGS, the California Geological Survey, and the Southern California Earthquake Center.

Sources: USGS Fact Sheet 2008-3018, "2008 United States National Seismic Hazard Maps," (April 2008), at [http://pubs.usgs.gov/fs/ 2008/3018/pdf/FS08-3018_508.pdf]; USGS Fact Sheet 2008-3027, "Forecasting California's Earthquakes — What Can We Expect in the Next 30 Years?" (2008), at [http://pubs.usgs.gov/fs/2008/3027/fs2008-3027.pdf].

Sources: USGS Fact Sheet 2008-3018, "2008 United States National Seismic Hazard Maps," (April 2008), at [http://pubs.usgs.gov/fs/2008/30 1 8/pdf/FS08-30 1 8_508.pdf]; USGS Fact Sheet 2008-3027, "Forecasting California's Earthquakes — What Can We Expect in the Next 30 Years?" (2008), at [http://pubs.usgs.gov/fs/2008/3027/fs2008- 3027.pdf].

**Table 1. 26 Urban Areas Facing Significant Seismic Risk
(alphabetically by state for cities with at least 300,000 people)**

State	City	State	City
Alaska	Anchorage	Nevada	Las Vegas
California	Fresno	Nevada	Reno
California	Los Angeles	New Mexico	Albuquerque
California	Sacramento	New York	New York
California	Salinas	Oregon	Eugene-Springfield
California	San Diego	Oregon	Portland
California	San Francisco-Oakland	Puerto Rico	San Juan
California	Santa Barbara	South Carolina	Charleston
California	Stockton-Lodi	Tennessee	Chattanooga-Knoxville
Idaho	Boise	Tennessee	Memphis
Indiana	Evansville	Utah	Provo-Orem
Massachusetts	Boston	Utah	Salt Lake City
Missouri	St. Louis	Washington	Seattle

Sources: USGS Fact Sheet 2006-3016 (March 2006); USGS Circular 1188, Table 3.
Note: These areas are identified using a population-based risk factor based on 1999 population data. (William Leith, ANSS Coordinater, USGS, Reston, VA, telephone conversation, Nov. 15, 2006).

The USGS estimates that several million earthquakes occur worldwide each year, but the majority are of small magnitude or occur in remote areas, and are not detectable. More earthquakes are detected each year as more seismometers[2] are installed in the world, but the number of large earthquakes (magnitude greater than 6.0)[3] has remained relatively constant. Between 2000 and 2007 there were 2,261- 3,876 earthquakes per year in the United States, according to the National Earthquake Information Center (NEIC). (See Figure 2.)

As Figure 2 shows, about 98% of earthquakes detected each year by the NEIC are smaller than magnitude 5.0; only 55 earthquakes exceeded magnitude 6.0 for the eight-year period (less than 0.3% of the total earthquakes detected) for an average of slightly less than seven earthquakes per year of at least 6.0 magnitude.

Large earthquakes, although infrequent, cause the most damage and are responsible for most earthquake-related deaths. Over the past 100 years, relatively few Americans have died as a result of earthquakes, compared to citizens in other countries.[4] The great San Francisco earthquake of 1906 claimed an estimated 3,000 lives, as a result of both the earthquake and subsequent fires. Since 1970, three major earthquakes in the United States were responsible for 188 of the 212 total earthquake-related fatalities (see Table 2).

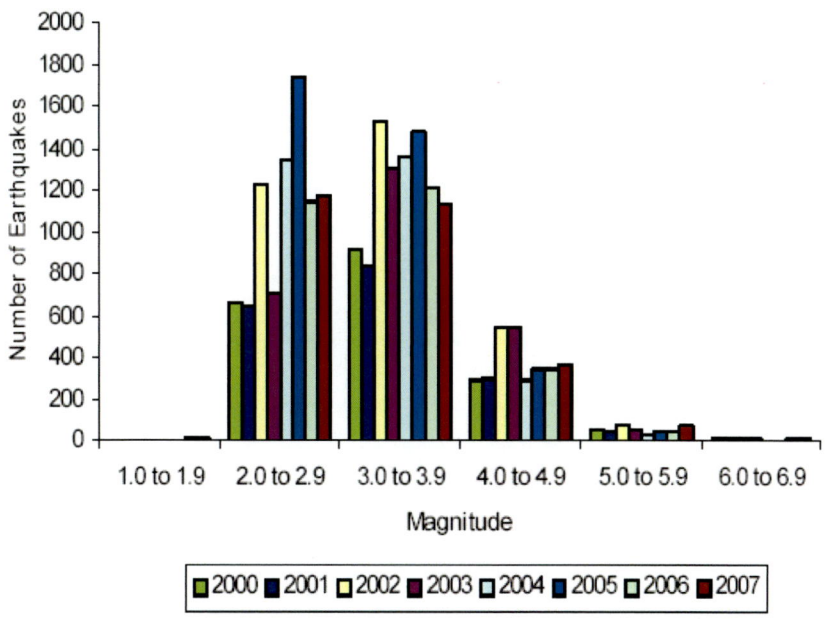

Source: USGS, "Earthquake Facts and Statistics," at [http://neic.usgs.gov/neis/eqlists/eqstats.html]; data as of June 10, 2008.
Note: Earthquakes greater than magnitude 7.0 and less than 1.0 are not shown.

Figure 2. Histogram of the Number of U.S. Earthquakes from 2000 to 2007 by Magnitude (1.0 to 6.9).

Table 2. Earthquakes Responsible for Most U.S. Fatalities Since 1970

Date	Location	Magnitude	Deaths
February 9, 1971	San Fernando Valley, CA	6.6	65
October 18, 1989	Loma Prieta, CA	6.9	63
January 17, 1994	Northridge, CA	6.7	60

Source: USGS, [http://earthquake.usgs.gov/regional/states/us_deaths.php].
Note: Other sources report different numbers of fatalities associated with the Northridge earthquake.

Since 2000, only two deaths directly caused by earthquakes have occurred in the United States, both associated with falling debris in Paso Robles (CA) during the December 22, 2003, San Simeon earthquake of magnitude 6.5. In contrast, earthquakes have been directly or indirectly responsible for more than 430,000 fatalities in other countries since 2000. More than half of those estimated deaths resulted from the December 2004 Indonesian earthquake of magnitude 9.1 and the resulting tsunami. On May 12, 2008, a magnitude 7.9 earthquake struck Eastern Sichuan, China, causing the known deaths of nearly 70,000 people (see box).

MAGNITUDE 7.9 EARTHQUAKE ON MAY 12, 2008, IN CHINA

On May 12, 2008, at 2:28 PM local time (2:28 AM eastern daylight time), a catastrophic earthquake of magnitude 7.9 struck Eastern Sichuan, China. The epicenter is located approximately 960 miles southwest of Beijing, and the earthquake was triggered approximately 12 miles below the Earth's surface. As of June 11, 2008, nearly 70,000 fatalities had been reported. The earthquake was felt in parts of eastern, southern, and central China, and as far away as Bangladesh, Taiwan, Thailand, and Vietnam. Several large aftershocks have occurred since the main seismic event.

The May 12 earthquake resulted from movement along a northeast-trending *reverse* or *thrust* fault, reflecting stresses resulting from the convergence of rocks of the Tibetan Plateau, to the west, against the crust underlying the Sichuan Basin and southeastern China. The region has experienced large earthquakes in the past; on August 25, 1933, a magnitude 7.5 earthquake struck the northwestern margin of the Sichuan Basin, resulting in approximately 9,300 fatalities.

Some concerns have been raised about the possibility of an earthquake of similar magnitude occurring in a seismically active region of the United States, such as southern California, where fault movement similar to the Eastern Sichuan earthquake may occur. On May 22, 2008, the USGS released a hypothetical scenario for a magnitude 7.8 southern California earthquake, called the ShakeOut Scenario. In the scenario, scientists hypothetically simulated the ground shaking and fault rupture associated with a magnitude 7.8 earthquake, and estimated the resulting damage to buildings and infrastructure. The scenario estimated approximately 1,800 fatalities and $213 billion in economic losses as a result of the earthquake. The report points to aggressive retrofitting programs that have increased the seismic resistance of buildings, highways, and other critical infrastructure in southern California as one reason why the number of possible fatalities is relatively low.

> Some scientists have raised the possibility that earthquakes, such as the May 12 Sichuan event, may sometimes exhibit cascading behavior, where bursts of seismic energy are released along different places in a single fault, or jump between connected faults. Earthquakes that occur along the Sierra Madre Fault in southern California, for example, could trigger a series of cascading seismic events along other faults, such as the San Andreas. Seismic hazards estimates may not fully account for the damage that could be caused by cascading earthquakes along a connected fault system. Scientists are hoping to examine the Sichuan earthquake in more detail to better understand the nature of cascading seismic events and how they affect the U.S. seismic hazard estimates.
>
> **Sources**: Ken Hudnut, geophysicist, USGS, Pasadena, CA, phone conversation, June 11, 2008; USGS Earthquake Hazards Program, at [http://earthquake.usgs.gov/eqcenter/eqinthenews/2008/us2008ryan/#summary] and [http://earthquake.usgs.gov/eqcenter/]. eqinthenews/2008/us2008ryan/]; and USGS, The ShakeOut Scenario, Open-File Report 2008-1150 (2008), at [http://pubs.usgs.gov/of/2008/1 150/].]

The 1994 Northridge earthquake was the nation's most damaging earthquake in the past 100 years, preceded five years earlier by the second most costly earthquake — Loma Prieta. Table 3 shows the 10 costliest U.S. earthquakes in terms of insured and uninsured losses. Comparing losses between different earthquakes, and between earthquakes and other disasters such as hurricanes, can be difficult because of the different ways losses are calculated. Calculations may include a combination of insured losses, uninsured losses, and estimates of lost economic activity. For example, insured losses from Hurricane Katrina in 2005 — mainly property — may be $41 billion, according to one estimate.[5] Total property damage would rise if uninsured property were included; and including interrupted economic activity in the calculation could bring the total loss for Hurricane Katrina to $100 billion, according to one estimate.[6]

The United States faces potentially large total losses due to earthquake-caused damage to buildings and infrastructure and lost economic activity. As urban development continues in earthquake-prone regions in the United States, concerns are increasing about the exposure of the built environment, including utilities and transportation systems, to potential earthquake damage.[7] One estimate of loss from a severe earthquake in the Los Angeles area is over $500 billion. An even higher estimate — approximately $900 billion — includes damage to the heavily populated central New Jersey-Philadelphia corridor if a 6.5 magnitude earthquake occurred along a fault lying between New York City and Philadelphia.[8]

Table 3. The 10 Most Damaging Earthquakes in the United States

Year	Location	Magnitude	2005 constant $
1994	Northridge, CA	6.7	$26 billion[a]
1989	Loma Prieta, CA	6.9	$11 billion
1964	Anchorage, AK	9.2	$3.1 billion
1971	San Fernando, CA	6.5	$2.7 billion
2001	Nisqually, WA	6.8	$2.5 billion
1987	Whittier Narrows, CA	5.9	$615 million
1933	Long Beach, CA	6.3	$600 million
1953	Kern County, CA	7.5	$440 million
1992	Landers, CA	7.6	$130 million
1992	Cape Mendocino, CA	7.1	$92 million

Source: Insurance Information Institute, at [http://www.iii.org/media/facts/ stats byissue/ earthquakes/].

Note: Includes insured and uninsured losses.

[a.] Estimates for total losses resulting from the Northridge earthquake vary; the Congressional Budget Office estimated $43 billion in total losses ($50 million in 2005 dollars). See *Federal Reinsurance for Disasters*, Congressional Budget Office (September 2002), p. 19.

Some studies and techniques combine seismic risk with the value of the building inventory[9] and income losses (e.g., business interruption, wage, and rental income losses) in cities, counties, or regions across the country to provide estimations of economic losses from earthquakes. One report[10] calculates that the *annualized* loss from earthquakes nationwide is $4.4 billion, with California, Oregon, and Washington accounting for $3.7 billion (84%) of the U.S. total estimated annualized loss. Table 4 shows cities with estimated annualized U.S. earthquake losses over $10 million. Annualized earthquake loss (AEL) addresses two components of seismic risk: the probability of ground motion and the consequences of ground motion. It enables comparison between different regions with different seismic hazards and different building construction types and quality. For example, earthquake hazard is higher in the Los Angeles area than in Memphis, but the general building stock in Los Angeles is more resistant to the effects of earthquakes. The AEL annualizes the expected losses by averaging them by year.

A single large earthquake can cause far more damage than the average annual estimate. However, annualized estimates help provide comparisons of infrequent, high impact events like damaging earthquakes, with more frequently occurring hazards like floods, hurricanes, or other types of severe weather. The annualized earthquake loss values shown in Table 4 represent future estimates, and are calculated by multiplying losses from all potential future ground motions by their respective frequencies of occurrence, and then summing these values.[11]

Table 4. U.S. Cities With Estimated Annualized Earthquake Losses of More than $10 Million (in millions)

Rank	Metro area	AEL	Rank	Metro area	AEL
1	Los Angeles, CA	$1,069	21	Bakersfield, CA	$31
2	Riverside, CA	$357	22	Tacoma, WA	$28
3	Oakland, CA	$349	23	Las Vegas, NV	$28
4	San Francisco, CA	$346	24	Anchorage, AK	$25
5	San Jose, CA	$243	25	Boston, MA	$23
6	Orange, CA	$214	26	Hilo, HI	$20
7	Seattle, WA	$128	27	Stockton, CA	$19
8	San Diego, CA	$128	28	Reno, NV	$18
9	Portland, OR	$98	29	Memphis, TN	$17
10	Ventura, CA	$89	30	Philadelphia, PA	$17
11	New York, NY	$56	31	San Luis Obispo, CA	$16
12	Vallejo, CA	$53	32	Salem, OR	$15
13	Santa Rosa, CA	$51	33	Fresno, CA	$14
14	Salt Lake City, UT	$40	34	Charleston, SC	$13
15	Sacramento, CA	$39	35	Albuquerque, NM	$13
16	St. Louis, MO	$34	36	Newark, NJ	$12
17	Eureka, CA	$34	37	Honolulu, HI	$12
18	Salinas, CA	$33	38	Atlanta, GA	$11
19	Santa Barbara, CA	$33	39	Modesto, CA	$11
20	Santa Cruz, CA	$33	40	Redding, CA	$10

Source: FEMA Publication 366, *HAZUS 99 Estimated Annualized Earthquake Losses for the United States* (February 2001). Annualized earthquake losses (AEL) calculated in 2000 dollars.

Estimating earthquake damage is not an exact science and depends on many factors. Primarily, these are the probability of ground motion occurring in a particular area (see Figure 1), and the consequences of that ground motion, which are largely a function of building construction type and quality, and of the level of ground motion and shaking during the actual event. Some researchers have questioned whether the probability of ground motion estimates for regions of the country that experience infrequent earthquakes, such as the New Madrid Seismic Zone, are too high.[12] These researchers question whether the benefits of building structures to conform with the earthquake probability estimates merit the costs, in light of the uncertainty in making those probability estimates.[13] An uncertainty analysis of the seismic hazard in the New Madrid Seismic Zone is beyond the scope of this report.

The New Madrid Seismic Zone in the central United States is vulnerable to large but infrequent earthquakes. A series of large (magnitude greater than 7.0) earthquakes struck the Mississippi Valley over the winter of 1811-1812, centered close to the town of New Madrid, MO. Some of the tremors were felt as far away as Charleston, SC, and Washington, DC. The mechanism for the earthquakes in the New Madrid zone is poorly understood,[14] and no earthquakes of comparable magnitude have occurred in the area since these events. Such factors contribute to the difficulty of making a reasonable damage estimate for a low-frequency, high-impact event in the region based on the probability of an earthquake of similar magnitude occurring. This uncertainty has implications for policy decisions to ameliorate risk, such as setting building codes, and for designing and building structures to withstand a level of shaking commensurate with the risk. Developers of building codes tend to err on the side of caution.

Table 4 also shows annualized earthquake losses for the cities of New York, Boston, and Newark, where no destructive earthquakes have struck for generations.[15] Those cities represent areas of relatively low seismic hazard, but have high populations and dense infrastructure, which produces a significant risk to people and structures, according to some estimates.[16] In the absence of any significant or damaging earthquakes for those cities in recent memory, however, the actual risk is difficult to grasp intuitively.

MONITORING

Congress authorized the USGS to monitor seismic activity in the United States in the 1990 reauthorization of the National Earthquake Hazards Reduction Act (P.L. 101-614). The USGS operates a nationwide network of seismographic stations called the Advanced National Seismic System (ANSS), which includes the National Strong- Motion Project (NSMP). Globally, the USGS and the Incorporated Research Institutions for Seismology (IRIS) operate 140 seismic stations of the Global Seismic Network (GSN) in more than 80 countries. The GSN provides worldwide coverage of earthquakes, including reporting and research, and also monitors nuclear explosions.

Advanced National Seismic System (ANSS)

"The mission of ANSS is to provide accurate and timely data and information products for seismic events, including their effects on buildings and structures, employing modern monitoring methods and technologies."[17] If fully implemented, ANSS would encompass more than 7,000 earthquake sensor systems covering parts of the nation vulnerable to earthquake hazards. The system includes over 700 stations consisting of backbone stations, dense urban networks, and regional networks.[18] In the original conception for ANSS, approximately 6,000 of the planned stations are to be installed in 26 high- risk urban areas to monitor strong ground shaking and how buildings and other structures respond. Currently, five high-risk urban areas have instruments deployed in sufficient density to generate the data to produce near real-time maps, called ShakeMaps,[19] which can be used in emergency response during and after an earthquake.

Approximately 1,000 new instruments are to replace aging and obsolete stations in the networks that now monitor the nation's most seismically active regions. The current regional networks contain a mix of modern, digital, broadband, and high- resolution instruments that can provide real-time data; they are supplemented by older instruments that may require manual downloading of data. Universities in the region typically operate the regional networks and will likely continue to do so as ANSS is implemented.

Lastly, approximately 100 instruments comprise the existing "backbone" of ANSS, with a roughly uniform distribution across the United States, including Alaska and Hawaii. These instruments provide a broad and uniform minimum threshold of coverage across the country. The backbone network consists of USGSdeployed instruments and other instruments that serve both ANSS and the EarthScope project (described below, under "Research — Understanding Earthquakes").

In 2004, Congress passed the National Earthquake Hazards Reduction Program Reauthorization Act of 2004 (P.L. 108-360), which authorized $30 million for ANSS in FY2005 and $36 million per year through FY2009. Congress first authorized the program with P.L. 106-503 at a level of $38 million for FY2002 and $44 million for FY2003. Total expenditures for ANSS from FY2002 to FY2007 are slightly more than $37 million, or approximately 17% of authorized levels. The FY2009 budget request states that the USGS plans to install a total of 803 ANSS monitoring stations by the end of 2008, with no new stations planned after that. That would represent slightly more than 11% of the 7,000 seismic stations originally envisioned for the program.

National Strong-Motion Project (NSMP)

Under ANSS, the USGS operates the NSMP to record seismic data from damaging earthquakes in the United States on the ground and in buildings and other structures in densely urbanized areas. The program currently has 900 strong-motion[20] instruments in 701 permanent stations across the United States and in the Caribbean. The NSMP has three components: data acquisition, data management, and research. The near real-time measurements collected by the NSMP are used by other government agencies for emergency response and real-time warnings. If fully implemented, the ANSS program would deploy about 3,000 strong-motion instruments, and the NSMP program would operate those strong-motion instruments located in buildings and other structures. Many of the current NSMP instruments are older designs and are being upgraded with modern seismometers.

Global Seismic Network (GSN)

The GSN is a system of broadband digital seismographs arrayed around the globe and designed to collect high-quality data that are readily accessible to users worldwide, typically via computer. Currently, 140 stations have been installed in 80 countries and the system is nearly complete, although in some regions the spacing and location of stations has not fully met the original goal of uniform spacing of approximately 2,000 kilometers. The system is currently providing data to the United States and other countries and institutions for earthquake reporting and research, and for monitoring nuclear explosions to assess compliance with the Comprehensive Test Ban Treaty. The Emergency Supplemental Appropriations Act for Defense, the Global War on Terror, and Tsunami Relief, 2005 (P.L. 109-13) provided more than $8 million to the USGS, of which $1.45 million was to expand the GSN real-time communications.[21] Funding for the GSN totaled $7.3 million in FY2007.[22]

The Incorporated Research Institutions for Seismology (IRIS)[23] coordinates the GSN and manages and makes available the large amounts of data that are generated from the network. The actual network of seismographs is organized into two main components, each managed separately. The USGS operates two-thirds of the stations from its Albuquerque Seismological Laboratory, and the University of California-San Diego manages the other third via its Project IDA (International Deployment of Accelerometers). Other universities and affiliated agencies and institutions operate a small number of additional stations. IRIS, with funding from the NSF, supports all of the stations not funded through the USGS appropriations.

DETECTION, NOTIFICATION, AND WARNING

Unlike other natural hazards, such as hurricanes, where predicting the location and timing of landfall is becoming increasingly accurate, the scientific understanding of earthquakes does not yet allow for precise earthquake prediction. Instead, notification and warning typically involves communicating the location and magnitude of an earthquake as soon as possible after the event to emergency response providers and others who need the information.

Some probabilistic earthquake forecasts are being made available now that give, for example, a 24-hour probability of earthquake aftershocks for a particular region, such as California. These forecasts are not predictions, and are currently intended to increase public awareness of the seismic hazard, improve emergency response, and increase scientific understanding of the short-term hazard.[24] In the California example, a time-dependent map is created and updated every hour by a system that considers all earthquakes, large and small, detected by the California Integrated Seismic Network,[25] and calculates a probability that each earthquake will be followed by an aftershock[26] that can cause strong shaking. The probabilities are calculated from known behavior of aftershocks and the possible shaking pattern based on historical data.

When a destructive earthquake occurs in the United States or in other countries, the first reports of its location, or epicenter,[27] and magnitude originate either from the National Earthquake Information Center in Golden, CO, or from one of the regional seismic networks that are part of ANSS. Other organizations, such as universities, consortia, and individual seismologists may also contribute information about the earthquake after the event. Products, such as

ShakeMap, are assembled as rapidly as possible to assist in emergency response and damage estimation following a destructive earthquake.

National Earthquake Information Center (NEIC)

The NEIC, part of the USGS, is located in Golden, CO. Originally established as part of the National Ocean Survey (Department of Commerce) in 1966, the NEIC was made part of the USGS in 1973. With data gathered from the networks described above and from other sources, the NEIC determines the location and size of all destructive earthquakes that occur worldwide and disseminates the information to the appropriate national or international agencies, government public information channels, news media, scientists and scientific groups, and the general public.

The NEIC has long-standing agreements with key emergency response groups, federal, state, and local authorities, and other key organizations in earthquake-prone regions who receive automated alerts — typically location and magnitude of an earthquake — within a few minutes of an event in the United States. The NEIC sends these preliminary alerts by email and pager immediately after an earthquake's magnitude and epicenter are automatically determined by computer.[28] This initial determination is then checked by around-the-clock staff who confirm and update the magnitude and location data.[29] After the confirmation, a second set of notifications and confirmations are triggered to key recipients by email, pager, fax, and telephone.

For earthquakes outside the United States, the NEIC notifies the State Department Operations Center, and often sends alerts directly to staff at American embassies and consulates in the affected countries, to the International Red Cross, the U.N. Department of Humanitarian Affairs, and other recipients who have made arrangements to receive alerts.

With the advent of the USGS Earthquake Notification Service (ENS), notifications of earthquakes detected by the ANSS/NEIC are provided free to interested parties. Users of the service can specify the regions of interest, establish notification thresholds of earthquake magnitude, designate whether they wish to receive notification of aftershocks, and even set different magnitude thresholds for daytime or nighttime to trigger a notification.

ShakeMap

Traditionally, the information commonly available following a destructive earthquake has been epicenter and magnitude, as in the data provided by the NEIC described above. Those two parameters by themselves, however, do not always indicate the intensity of shaking and extent of damage following a major earthquake. Recently, the USGS developed a product called ShakeMap that provides a near real-time map of ground motion and shaking intensity following an earthquake in areas of the United States where the ShakeMap system is in place. Currently, ShakeMaps are available for northern California, southern California, the Pacific Northwest, Nevada, Utah, and Alaska.[30] Figure 3 shows an example of a ShakeMap.

With improvements to the regional seismographic networks in the areas where ShakeMap is available, new real-time telemetry from the region, and advances in digital communication and computation, ShakeMaps are now triggered automatically and made available within minutes of the event via the Web. In addition, better maps are now available because of recent improvements in understanding the relationship between the ground motions recorded during the earthquake and the intensity of resulting damage. The maps produced portray the extent of damaging shaking and can be used by emergency response and for estimating loss following a major earthquake. If databases containing inventories of buildings and lifelines[31] are available, they can be combined with shaking intensity data to produce maps of estimated damage.

The ShakeMaps have limitations, especially during the first few minutes following an earthquake before more data arrive from distributed sources. Because they are generated automatically, the initial maps are preliminary, and may not have been checked by human oversight when first made available. They are considered a work in progress, but are deemed to be very promising, especially as more modern seismic instruments are added to the regional networks under ANSS and the computational and telecommunication ability improves.

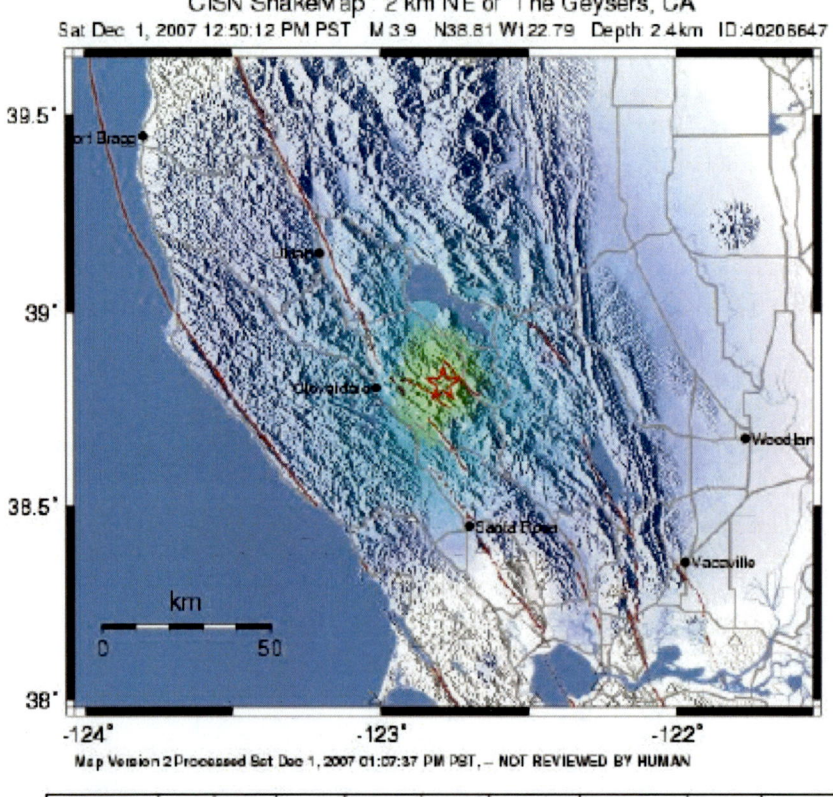

Source: USGS, [http://earthquake.usgs.gov/eqcenter/shakemap/nc/shake/40206647/].
Note: Earthquake occurred northeast of the The Geysers, CA, on December 1, 2007, at 12:50 p.m., with a magnitude of 3.9. Viewed on January 12, 2008.

Figure 3. Example of a ShakeMap.

NATIONAL EARTHQUAKE HAZARDS REDUCTION PROGRAM (NEHRP)

In 1977 Congress passed the Earthquake Hazards Reduction Act (P.L. 95-124) establishing NEHRP as a long-term earthquake risk reduction program for the United States. The program initially focused on research, led by USGS and NSF, toward understanding and ultimately predicting earthquakes. Earthquake prediction has proved intractable thus far, and the NEHRP program shifted its focus to minimizing losses from earthquakes after they occur. FEMA was created in 1979 and President Carter designated it as the lead agency for NEHRP. In 1980, Congress reauthorized the Earthquake Hazards Reduction Act (P.L. 96-472), defining FEMA as the lead agency and authorizing additional funding for earthquake hazard preparedness and mitigation to FEMA and the National Bureau of Standards (now NIST).

Mitigation

In 1990, Congress reauthorized NEHRP (P.L. 101-614) and made substantive changes, to decrease the emphasis on earthquake prediction, clarify the role of FEMA, clarify and expand the program objectives, and require federal agencies to adopt seismic safety standards for new and existing federal buildings. In 2004, Congress reauthorized NEHRP through FY2009 (P.L. 108-3 60) and shifted primary responsibility for planning and coordinating NEHRP from FEMA to NIST. It also established a new interagency coordinating committee and a new advisory committee, both focused on earthquake hazards reduction.

The current program activities are focused on four broad areas:

- developing effective measures to reduce earthquake hazards;
- promoting the adoption of earthquake hazards reduction measures by federal, state, and local governments, national building standards and model building code organizations, engineers, architects, building owners, and others who play a role in planning and constructing buildings, bridges, structures, and critical infrastructure or lifelines;
- improving the basic understanding of earthquakes and their effects on people and infrastructure, through interdisciplinary research involving engineering, natural sciences, and social, economic, and decision sciences; and

- developing and maintaining ANSS, the George E. Brown Jr. Network for Earthquake Engineering Simulation (NEES),[32] and the GSN.

The House Science Committee report in the 108th Congress on H.R. 2608 (P.L. 108-360) noted that NEHRP has produced a wealth of useful information since 1977, but it also stated that the program's potential has been limited by the inability of the NEHRP agencies to coordinate their efforts.[33] The committee asserted that restructuring the program with NIST as the lead agency, directing funding towards appropriate priorities, and implementing it as a true interagency program would lead to improvement.

Under the reauthorization, the Director of NIST chairs the Interagency Coordinating Committee, which is composed of the directors of FEMA, USGS, NSF, the Office of Science and Technology Policy, and the Office of Management and Budget. The Interagency Coordinating Committee is charged with overseeing the planning, management, and coordination of the program. Primary responsibilities for the NEHRP agencies break down as follows:

- NIST supports the development of performance-based seismic engineering tools and works with other groups to promote the commercial application of the tools through building codes, standards, and construction practices;
- FEMA assists other agencies and private-sector groups to prepare and disseminate building codes and practices for structures and lifelines, and aid development of performance-based codes for buildings and other structures;
- USGS conducts research and other activities to characterize and assess earthquake risks, and (1) operates a forum, using the NEIC, for the international exchange of earthquake information; (2) works with other NEHRP agencies to coordinate activities with earthquake reduction efforts in other countries; and (3) maintains seismic hazard maps in support of building codes for structures and lifelines, and other maps needed for performance-based design approaches; and
- NSF supports research to improve safety and performance of buildings, structures, and lifelines using the large-scale experimental and computational facilities of NEES and other institutions engaged in research and implementation of NEHRP.

Table 5 shows authorization of appropriations for NEHRP from FY2005 through FY2008 and the enacted amounts by agency through FY2007. The total enacted amount for FY2005-FY2007 is $366.8 million, $157.2 million less than the amount authorized in P.L. 108-360 of $524 million over the three-year span. Slightly less than $200 million is authorized under the law for FY2009.

Table 5. Authorized and Enacted Funding for NEHRP ($ millions)

Agency	FY05 Auth.	FY05 Enact.	FY06 Auth.	FY06 Enact.	FY07 Auth.	FY07 Enact.	FY08 Auth.	FY08 Enact.	FY09 Auth.
NIST	10.0	0.9	11.0	0.9	12.1	1.7	13.3	1.7	14.64
FEMA	21.0	14.7	21.6	9.5	22.3	9.1	23.0	6.1	23.64
USGS	77.0	58.4	84.4	54.5	85.9	55.4	87.4	58.1	88.9
NSF	58.0	53.1	59.5	53.8	61.2	54.8	62.9	55.6	64.7
Total	*166.0*	*127.1*	*176.5*	*118.7*	*181.5*	*121.0*	*186.6*	*121.5*	*191.8*
									8

Source: NEHRP program office, at [http://www.nehrp.gov/plans/index.htm].

HAZUS-MH

FEMA, under contract with the National Institute of Building Sciences,[34] developed a methodology and software program called the Hazards U.S. Multi-Hazard (HAZUS-MH).[35] The program allows a user to estimate losses from damaging earthquakes, hurricane winds, and floods before a disaster occurs. The pre-disaster estimates could provide a basis for developing mitigation plans and policies, preparing for emergencies, and planning response and recovery. HAZUSMH combines existing scientific knowledge about earthquakes (for example, ShakeMaps, described above), engineering information that includes data on how structures respond to shaking, and geographic information system (GIS) software to produce maps and display hazards data including economic loss estimates. The loss estimates produced by HAZUS-MH include the following:

- physical damage to residential and commercial buildings, schools, critical facilities, and infrastructure;
- economic loss, including lost jobs, business interruptions, repair and reconstruction costs; and
- social impacts, including estimates of shelter requirements, displaced households, and number of people exposed to the disaster.

In addition to furnishing information as part of earthquake mitigation efforts, HAZUS-MH can also be used to support real-time emergency response activities by state and federal agencies after a disaster. Twenty-seven HAZUS-MH user groups[36] — cooperative ventures among private, public, and academic organizations that use the HAZUS-MH software — have formed across the United States to help foster better-informed risk management for earthquakes and other natural hazards. HAZUS-MH software was first released to the public in 1997 and the first user group, the Bay Area HAZUS-MH User Group, was formed the same year.

Research — Understanding Earthquakes

U.S. Geological Survey
Under NEHRP, the USGS has responsibility for conducting targeted research into improving the basic scientific understanding of earthquake processes. The current earthquake research program at the USGS covers six broad categories:[37]

- *Borehole geophysics and rock mechanics*: studies to understand heat flow, stress, fluid pressure, and the mechanical behavior of fault-zone materials at seismogenic[38] depths to yield improved models of the earthquake cycle;
- *Crustal deformation*: studies of the distortion or deformation of the earth's surface near active faults as a result of the motion of tectonic plates;
- *Earthquake geology and paleoseismology*: studies of the history, effects, and mechanics of earthquakes;
- *Earthquake hazards*: studies of where, why, when, and how earthquakes occur;
- *Regional and whole-earth structure*: studies using seismic waves from earthquakes and man-made sources to determine the structure of the planet ranging from the local scale, to the whole crust, mantle, and even the earth's core; and
- *Strong-motion seismology, site response, and ground motion*: studies of large-amplitude ground motions and the response of engineered structures to those motions using accelerometers.

National Science Foundation

NSF supports fundamental research into understanding the earth's dynamic crust. Through its Earth Sciences Division[39] (part of the Geosciences Directorate), NSF distributes research grants and coordinates programs investigating the crustal processes that lead to earthquakes around the globe. Recently, NSF initiated a Major Research Equipment and Facilities Construction (MREFC) project called EarthScope.[40] EarthScope is deploying instruments across the United States to study the structure and evolution of the North American Continent, and to investigate the physical processes that cause earthquakes and volcanic eruptions. EarthScope, a five-year, $200 million project, began in 2003, is funded by NSF, and is conducted in partnership with the USGS and NASA.

EarthScope instruments will form a framework for broad, integrated studies of the four-dimensional (three spatial dimensions, plus time) structure of North America. The project is divided into three main programs:

- *The San Andreas Fault Observatory at Depth (SAFOD)*: a deep borehole observatory drilled through the San Andreas fault zone close to the hypocenter of the 1966 Parkfield, CA, magnitude 6 earthquake.
- *The Plate Boundary Observatory (PBO)*: a system of GPS arrays and strainmeters[41] that measure the active boundary zone between the Pacific and North American tectonic plates in the western United States.
- *USArray*: four hundred transportable seismometers that will be deployed systematically across the United States on a uniform grid to provide a complete image of North America from continuous seismic measurements.

Through its Engineering Directorate, NSF funds NEES,[42] a project intended to operate until 2014, aimed at understanding the effects of earthquakes on structures and materials. To achieve the program's goal, the facilities conduct experiments and computer simulations of how buildings, bridges, utilities, coastal regions, and materials behave during an earthquake.

CONCLUSIONS

The 2003 reauthorization of NEHRP shifted leadership of the multiagency program from FEMA to NIST and authorized the program through FY2009.

Congress may wish to determine whether the reorganized structure has yielded expected benefits for the program, now in its fourth year since P.L. 108-3 60 was enacted. Appropriations for NEHRP have not met levels authorized in the law for FY2005 through FY2008, falling short by an average of 31% for FY2005 through FY2008. What effect funding at the levels enacted through FY2008 has had on all programs authorized under NEHRP is unclear, although progress in some programs has been slower than anticipated. For example, the Advanced National Seismic System (ANSS), an integrated system of earthquake sensors deployed across the country, is about 11% complete, in part because appropriated funds for ANSS have historically been a fraction of the authorized levels. To what extent the current rate of progress toward meeting the goals of ANSS affects the U.S. capability to detect earthquakes and minimize losses after an earthquake occurs is not clear.

Congress may also wish to examine if and how new research results — generated under EarthScope, a research program at NSF distinct from NEHRP — on understanding earthquakes are contributing to the nation's resilience to earthquake disasters. An ongoing question is how scientific advances in understanding the fundamental nature of earthquakes — funded by EarthScope and other programs at NSF and the USGS — can be applied towards improving the U.S. capability to minimize losses from destructive earthquakes.

ADDITIONAL READING

Aspects of the federal role in the aftermath of a damaging earthquake or other natural catastrophes — the response and recovery phase — are covered in the following CRS reports.

CRS Report RL33330, Community Development Block Grant Funds in Disaster Relief and Recovery, by Eugene Boyd.
CRS Report RL33060, Tax Deductions for Catastrophic Risk Insurance Reserves: Explanation and Economic Risk Analysis, by David L. Brumbaugh and Rawle O. King.
CRS Report RL31734, Federal Disaster Recovery Programs: Brief Summaries, by Mary B. Jordan.
CRS Report RL32847, Tsunamis and Earthquakes: Is Federal Disaster Insurance in Our Future?, by Rawle O. King.

CRS Report RS22268, Repairing and Reconstructing Disaster-Damaged Roads and Bridges: The Role of Federal-Aid Highway Assistance, by Robert S. Kirk.

CRS Report RS22273, Emergency Contracting Authorities, by John R. Luckey.

CRS Report RL34 146, FEMA's Disaster Declaration Process: A Primer, by Francis X. McCarthy.

CRS Report RL33206, Vulnerability of Concentrated Critical Infrastructure: Background and Policy Options, by Paul W. Parfomak.

In response to conference report H.Rept. 109-699, accompanying the FY2007 Department of Homeland Security Appropriations Act (P.L. 109-295), FEMA released the following report to Congress:

Federal Emergency Management Agency, "Federal Earthquake Response Plans: Report to Congress," December 2007, 92 pages.

REFERENCES

[1] Magnitude is a number that characterizes the relative size of an earthquake. Earthquake magnitude is often reported using the *Richter* scale (magnitudes in this report are generally consistent with the Richter scale). Richter magnitude is calculated from the strongest seismic wave recorded from the earthquake, and is based on a logarithmic (base 10) scale: for each whole number increase in the Richter scale, the ground motion increases by ten times. The amount of energy released per whole number increase, however, goes up by a factor of 32. The *moment magnitude* scale is another expression of earthquake size, or energy released during an earthquake, that roughly corresponds to the Richter magnitude and is used by most seismologists because it more accurately describes the size of very large earthquakes. *Intensity* is a measure of how much shaking occurred at a site based on observations and amount of damage. Intensity is usually reported on the Modified Mercalli Intensity Scale as a Roman numeral ranging from I (not felt) to XII (total destruction).

[2] *Seismometers* are instruments that measure and record the size and force of seismic waves, essentially sound waves radiated from the earthquake as it ruptures. Seismometers generally consist of a mass attached to a fixed base. During an earthquake, the base moves and the mass does not, and the relative motion is commonly transformed into an electrical voltage that is recorded. A *seismograph* usually refers to the *seismometer* and the recording device, but the two terms are often used interchangeably.

[3] See USGS "Earthquakes Facts and Statistics" at [http://neic.usgs.gov/neis/eqlists/eqstats. html#table_2].
[4] Estimates of earthquake-related fatalities vary and an exact tally of deaths and injuries is rare. For more information on the difficulties of counting earthquake-related deaths and injuries, see [http://earthquake.usgs.gov/regional/world/casualty_totals.php].
[5] Insurance Information Institute, [http://www.iii.org/media/facts/ statsby issue/ hurricanes/]. Loss estimates are in 2005 dollars.
[6] Risk Management Solutions (RMS), Newark, CA, press release (Sept. 2, 2005), at [http://www.rms.com/NewsPress/PR _090205 _HUKatrina.asp].
[7] FEMA Publication 366, *HAZUS 99 Estimated Annualized Earthquake Losses for the United States* (February 2001). Hereafter referred to as FEMA 366.
[8] A. M. Best Company Inc., *2006 Annual Earthquake Study: $100 Billion of Insured Loss in 40 Seconds* (Oldwick, NJ: A.M. Best Company, 2006), p. 12. The A. M. Best report includes estimates from catastrophe-modeling companies of predicted damage from hypothetical earthquakes in Los Angeles, the Midwest, the Northeast, and Japan. The report cites an estimate by one such company, Risk Management Solutions, that a hypothetical 7.4 magnitude event along the Newport-Inglewood Fault near Los Angeles would cause $549 billion in total property damage. A hypothetical 6.5 magnitude earthquake along a fault between Philadelphia and New York City would produce $901 billion in total loss, according to an RMS estimate.
[9] Building inventory refers to four main inventory groups: (1) general building stock, (2) essential and high potential loss facilities, (3) transportation systems, and (4) utility systems (FEMA 366).
[10] FEMA 366.
[11] FEMA 366
[12] Andrew Newman, Seth Stein, John Weber, Joseph Engeln, Ailin Mao, and Timothy Dixon, "Slow Deformation and Lower Seismic Hazard in the New Madrid Seismic Zone," *Science*, v. 284 (April 23, 1999), pp. 619-621.
[13] Seth Stein, Joseph Tomasello, and Andrew Newman, "Should Memphis Build for California's Earthquakes?", *Eos*, v. 84, no. 19, (May 13, 2003), pp. 177, 184-185.
[14] In contrast to California, where earthquakes occur on the active margin of the North American tectonic plate, the New Madrid seismic zone is not on a plate boundary but may be related to old faults in the interior of the plate, marking a zone of tectonic weakness.

[15] The largest earthquakes in New York, New Jersey, and Massachusetts were, respectively, 1944, Massena, NY, magnitude 5.8, felt from Canada south to Maryland; 1783, New Jersey, magnitude 5.3, felt from New Hampshire to Pennsylvania; 1755, Cape Ann and Boston, MA, intensity of VIII on the Modified Mercalli Scale, felt from Nova Scotia to Chesapeake Bay (USGS Earthquake Hazards Program).
[16] FEMA 366 and USGS Circular 1188, Table 3.
[17] USGS Earthquake Hazards Program, at [http://earthquake.usgs.gov /research/ monitoring/ anss/].
[18] USGS FY2009 Budget Justification, at [http://www.doi.gov/budget/ 2009/data/greenbook/FY2009_USGS_Greenbook.pdf].
[19] ShakeMap is a product of the USGS Earthquake Hazards Program in conjunction with regional seismic network operators; see Shakemap below.
[20] Strong motion seismometers, or accelerometers, are special sensors that measure the acceleration of the ground during large (>6.0 magnitude) earthquakes.
[21] See also CRS Report RL32739, *Tsunamis: Monitoring, Detection, and Early Warning Systems*, by Wayne A. Morrissey.
[22] William Leith, USGS, personal communication, Apr. 2, 2007.
[23] IRIS is a university research consortium, primarily funded by NSF, that collects and distributes seismographic data.
[24] USGS Open-File Report 2004-1390, and California 24-hour Aftershock Forecast Map, at [http://pasadena.wr.usgs.gov/step/].
[25] The California Integrated Seismic Network is the California region of ANSS; see [http://www.cisn.org/].
[26] Earthquakes typically occur in clusters, in which the earthquake with the largest magnitude is called the main shock, events before the main shock are called foreshocks, and those after are called aftershocks. See also [http://pasadena.wr.usgs.gov/step/aftershocks. html].
[27] The *epicenter* of an earthquake is the point on the Earth's surface directly above the hypocenter. The *hypocenter* is the location beneath the Earth's surface where the fault rupture begins.
[28] Stuart Simkin, NEIC, Golden, CO, telephone conversation, Nov. 4, 2006.
[29] In early 2006, the NEIC implemented an around-the-clock operation center and seismic event processing center in response to the Indonesian earthquake and resulting tsunami of December 2004. Funding to implement 24/7 operations was provided by P.L. 109-13.
[30] See [http://earthquake.usgs.gov/eqcenter/shakemap/].
[31] Lifelines are essential utility and transportation systems.

[32] NEES is an NSF-funded project that consists of 15 experimental facilities and an IT infrastructure with a goal of mitigating earthquake damage by the use of improved materials, designs, construction techniques, and monitoring tools.

[33] U.S. House, Committee on Science, *National Earthquake Hazards Reduction Program Reauthorization Act of 2003*, H.Rept. 108-246 (Aug. 14, 2003), p. 13.

[34] The National Institute of Building Sciences (NIBS) is a non-profit non-governmental organization established by Congress in the Housing and Community Development Act of 1974 (PL 99-383). NIBS is funded through dues from its membership, private sector contributions, and contracts with federal and state agencies. The mission of NIBS is to improve the building regulatory environment, facilitate introducing new products and technologies into the building process, and disseminate technical and regulatory information. See [http://www.nibs.org/].

[35] See [http://www.fema.gov/plan/prevent/hazus/hz_overview.shtm].

[36] See [http://www.hazus.org/].

[37] See [http://earthquake.usgs.gov/research/].

[38] Seismogenic means capable of generating earthquakes.

[39] See [http://www.nsf.gov/div/index.j sp?div=EAR].

[40] See [http://www.earthscope.org/].

[41] A strainmeter is a tool used by seismologists to measure the motion of one point relative to another.

[42] A non-profit NEES consortium (NEESinc.) has operated the facilities for the 10-year operating lifespan at the following institutions: Cornell University; Lehigh University; Oregon State University; Rensselaer Polytechnical Institute; University of Buffalo-State University of New York; University of California-Berkeley; University of California-Davis; University of California-Los Angeles; University of California-San Diego; University of California-Santa Barbara; University of Colorado-Boulder; University of Illinois at Urbana- Champaign; University of Minnesota; University of Nevada-Reno; University of Texas at Austin. See [http://www.nees.org/].

In: Earthquakes: Risk, Monitoring and Research ISBN: 978-1-60692-648-2
Editor: Earl V. Leary © 2009 Nova Science Publishers, Inc.

Chapter 2

ANNUAL REPORT OF THE NATIONAL EARTHQUAKE HAZARDS REDUCTION PROGRAM[*]

FEMA, National Institute of Standards and Technology, National Science Foundation, and USGS

PREFACE

Of all natural hazards threatening the United States, earthquakes pose the greatest risk in terms of casualties and damage. According to a 2006 National Research Council (NRC) report, 42 states have regions with some degree of earthquake potential and 18 states have areas of high or very high seismicity. Over 75 million people live in urban areas with moderate to high earthquake risk. The NRC report also notes that the estimated value of structures in all states prone to earthquake damage is approximately $8.6 trillion (2003 dollars). Recent studies estimate losses for several specific earthquake scenarios:

- Repeat of the 1906 San Francisco earthquake: $90–120 billion.
- A magnitude 7.5 earthquake in the Salt Lake City area: $18 billion.
- A magnitude 6.7 earthquake on the Seattle fault: $33 billion.

---

[*] This is an edited, excerpted and augmented edition of a FEMA, National Institute of Standards and Technology, National Science Foundation, and USGS publication.

Although earthquakes are inevitable, earthquake disasters are not. The National Earthquake Hazards Reduction Program (NEHRP), authorized by the Earthquake Hazards Reduction Act of 1977 (Public Law 95-124), as amended by the National Earthquake Hazards Reduction Program Reauthorization (Public Law 108-360), seeks to mitigate earthquake losses in the United States by performing basic and applied research, developing cost-effective risk reduction measures, and actively promoting their implementation.

Over the past 30 years, NEHRP has helped the nation make significant advances in earthquake safety; however, major challenges remain. During this period, the population of the United States has increased from 200 million to 300 million; much of this population increase is in seismic areas. Many elements of the national infrastructure have aged without replacement, and have never been tested by strong earthquake shaking. In 2003, the American Society of Civil Engineers (ASCE) reported (www.asce.org/ reportcard/2005/) that 27.1 percent of the nation's bridges (160,570) were structurally deficient or functionally obsolete. ASCE further reported that even more bridges in urban areas (31.2 percent) were structurally deficient or functionally obsolete. These data indicate the condition of the nation's infrastructure, and are vividly illustrated to the public when periodic bridge collapses are reported in the press. Maintaining global competitiveness requires that practices to mitigate earthquake impacts in the United States, both in new construction and in its aging infrastructure, be affordable and demonstrably cost-effective to all levels of government and private interests. This same global economy exposes our national economy to earthquake risks in other countries, particularly around the Pacific Rim.

Thus, although NEHRP has made progress toward national earthquake safety, the pressures caused by changing demographics and economic conditions and priorities continue to require that NEHRP be viable, flexible, and forward looking. NEHRP must build on past experience but remain focused on future risks and on developing the practical solutions needed to reduce or eliminate these risks.

This report to Congress describes the activities and achievements of the NEHRP agencies and their partners during 2007 in building on past results and advancing earthquake safety nationwide.

EXECUTIVE SUMMARY

The Interagency Coordinating Committee (ICC) of the National Earthquake Hazards Reduction Program (NEHRP) presents the NEHRP 2007 Annual Report. This has been an active and productive year for NEHRP with significant progress

Annual Report of the National Earthquake Hazards Reduction Program 31

toward long/term goals and objectives for reducing losses from future earthquakes. As required by Public Law 108/360, this report describes these activities and progress, and gives program budgets for fiscal year (FY) 2008 and those proposed for FY 2009.

The four NEHRP agencies are the Federal Emergency Management Agency (FEMA), the National Institute of Standards and Technology (NIST), the National Science Foundation (NSF), and the U.S. Geological Survey (USGS). The ICC is composed of the Directors of the NEHRP agencies and the Directors of the White House Office of Science and Technology Policy and Office of Management and Budget. NIST serves as the NEHRP lead agency and its Director chairs the ICC.

Total enacted funding for NEHRP in FY 2008 is $121.5 million, compared to $121.0 million in FY 2007. Total requested funding for NEHRP activities in FY 2009 is $124.5 million.

The principal accomplishments of NEHRP in FY 2007 are summarized below:

> NEHRP Leadership: The year 2007 was the first year with all of the new statutory management and advisory bodies in place and functioning. The Advisory Committee on Earthquake Hazards Reduction, composed of external scientists and engineers, was fully staffed and conducted its first full meeting in May. In 2007, the ICC met three times, providing expert and high level recommendations and oversight to NEHRP in general and to the new Strategic Plan in particular. This plan will be completed in early 2008 and its goals, objectives, and outcomes will provide direction and milestones for NEHRP during the next 5 years.
>
> NSF Research Centers: At the end of 2007, three earthquake engineering research centers completed their 10/year period of NSF support. These centers have made major contributions to the development of performance/based seismic design, improved fundamental understanding of seismic performance of structures, and advanced technologies to help mitigate and respond to seismic events. All three centers will continue as research centers through various combinations of university, state, and private sector support, and other federal funding. The momentum provided to these centers through NSF support has made them self/sustaining, and will allow them to continue making contributions to NEHRP goals.
>
> George E. Brown, Jr. Network for Earthquake Engineering Simulation (NEES): The construction phase of NEES was completed in 2004 and it has developed into an active research partnership. As of September 2007, NSF has funded over 40 research projects to use the NEES facilities to study a wide spectrum of engineering topics related to soil and structure integrity and safety in earthquakes and tsunamis. As a part of NEES operations, the NEES Consortium, Inc. continues to push the forefront of experimental testing with the improvement of the experimental infrastructure through new telepresence and hybrid simulation tools.

Earthquake Safety in Building Design and Construction Standards and Codes: FEMA developed tools, and associated training material, for design professionals, engineers, building contractors, and decision/makers to address a number of existing challenges. Topics covered include seismic rehabilitation of existing buildings, updating the Homebuilders' Guide to Earthquake Resistant Design and Construction, and developing the Design Guidefor Improving Hospital Safety in Earthquakes, Floods, and High Winds. As a result of FEMA's success in influencing the Nation's model building codes, states and localities are increasingly adopting and enforcing earthquake/ resistant building codes based on these model codes. In FY 2007, there was a 30 percent increase in the number of jurisdictions in high/seismic regions that adopted a disaster/resistant building code. All 50 states have now adopted a national model building code, either in whole or in part.

Earthquake Monitoring: The USGS continued its deployment of the Advanced National Seismic System (ANSS) that integrates, modernizes, and expands seismic monitoring, data analysis, and earthquake notification in the United States. By the end of FY 2007, ANSS had grown to 784 stations, which is just over 10 percent of the planned 7,100/station network. USGS began issuing rapid estimates of population exposure to surface shaking to help assess the impact of an earthquake disaster within a few tens of minutes after a large earthquake.

Assessing Seismic Hazards: The USGS provided new national seismic hazard maps to the Building Seismic Safety Council, initiating the process for their consideration in the 2012 version of the International Building Code. For California, a new seismic forecast model was developed by the Working Group on California Earthquake Probabilities, a collaboration between university scientists, the State of California, and USGS. The USGS also released new urban seismic hazard maps for Seattle, the first probabilistic seismic hazard maps based on three/dimensional simulations of earthquake ground motions.

These are just a few of the NEHRP accomplishments during FY 2007 described in this report. Although much has been accomplished, much remains to be done. Work completed in FY 2007 will have applications immediately or in the near future in reducing earthquake risk. Work advanced in FY 2007 has laid a strong foundation for realizing similarly effective outcomes in future years.

1. Introduction

1.1. Legislative Overview

Congress first authorized the National Earthquake Hazards Reduction Program (NEHRP) in 1977 with the goal of reducing earthquake losses nationwide. Congress last reauthorized NEHRP in 2004. The four federal agencies with funding authorizations and legislatively mandated responsibilities for

NEHRP activities are the Federal Emergency Management Agency (FEMA), the National Institute of Standards and Technology (NIST), the National Science Foundation (NSF), and the U.S. Geological Survey (USGS).

The 2004 reauthorization of NEHRP (Public Law 108/360) requires the NEHRP Interagency Coordinating Committee (ICC) to submit an annual report at the time of the President's annual budget request to Congress. This report meets this requirement and is transmitted accordingly to the Committee on Science and Technology and the Committee on Natural Resources of the House of Representatives and the Committee on Commerce, Science, and Transportation of the Senate.

1.2. NEHRP Organization and Management

During the past 3 years, NEHRP has undergone considerable change in organization, with the designation of a new lead agency, NIST, and the establishment of management and oversight committees and an external advisory panel. The NEHRP annual report for fiscal year (FY) 2005 and FY 2006 presented a detailed review of these changes and the current status of NEHRP management.

The work of NEHRP is carried out by the four NEHRP agencies and the organizations they support. The roles of the NEHRP agencies are described below.

Federal Emergency Management Agency
FEMA is responsible for developing effective earthquake risk reduction tools and promoting their implementation, as well as supporting the development of disaster/resistant building codes and standards. FEMA NEHRP activities are led by the Risk Reduction Division, Mitigation Directorate, FEMA Headquarters, and through the FEMA Regions. Organizations that received FEMA support for NEHRP activities in FY 2007 include states, multi/state earthquake consortia, seismic experts from the academic and practitioner communities, and stakeholder groups that access a vast network of expertise in engineering, academics, and public policy.

National Institute of Standards and Technology
NIST serves as the NEHRP lead agency and, in addition, develops, evaluates, and tests earthquake/ resistant design and construction practices for implementation in building codes and engineering practice. NEHRP Directorate,

Secretariat, and applied research activities are conducted in the Building and Fire Research Laboratory of NIST.

National Science Foundation

NSF supports basic research and research facilities in earth sciences, engineering, and social, behavioral, and economic sciences relevant to the understanding of the causes and impacts of earthquakes and to developing practical measures to reduce their effects. NSF NEHRP-related support is carried out primarily through research grants to individual universities, university consortia, and other organizations. These grants are awarded primarily through the NSF Directorate for Engineering and the Directorate for Geosciences.

U.S. Geological Survey

The USGS operates and supports earthquake monitoring, data analysis, and notification facilities; provides earthquake hazard assessments; and conducts and supports targeted research on earthquake causes and effects. The Earthquake Hazards Program Office at the USGS Headquarters leads the USGS NEHRP work. USGS research and monitoring activities are conducted by USGS scientists at offices in Albuquerque, New Mexico; Anchorage, Alaska; Golden, Colorado; Memphis, Tennessee; Menlo Park and Pasadena, California; and Seattle, Washington, as well as through grants and cooperative agreements with universities, state geological surveys, and other organizations.

Cooperating Organizations

NEHRP agencies support and work with many cooperating organizations, which are described briefly in Appendix A of this report. These organizations are essential in furthering the work of NEHRP in research, development, and implementation. Many of these organizations receive support from multiple NEHRP agencies and other sources with interests common to NEHRP goals.

1.3. NEHRP Coordination and Oversight

NEHRP is coordinated at the highest levels by the principals of the four agencies and at the working level by program officials directly responsible for program execution. NEHRP also is reviewed and guided by an external advisory panel of nongovernment experts.

Interagency Coordinating Committee

Congress established the ICC in 2004 to "...oversee the planning, management, and coordination of the Program." The ICC is composed of the Director/Administrator of each NEHRP agency as well as the Directors of the White House Office of Science and Technology Policy and Office of Management and Budget. In addition to program oversight, the ICC is responsible for developing the NEHRP Strategic Plan, an implementation plan, an integrated NEHRP budget, and annual reports. The Director of NIST chairs the ICC.

The ICC met in October 2006, May 2007, and September 2007 to address program coordination and policy issues. In 2007, the ICC released its first annual report to Congress as required under the 2004 NEHRP reauthorization. The ICC has developed and endorsed a process of unified interagency program planning with coordinated budget requests (to take effect with the FY 2010 budget); and approved a framework for a revised NEHRP Strategic Plan.

Advisory Committee on Earthquake Hazards Reduction

Congress established the Advisory Committee on Earthquake Hazards Reduction (ACEHR) in 2004 to assess:

- Trends and developments in science and engineering of earthquake hazard reduction. Effectiveness of NEHRP in carrying out specified activities.
- The need to revise NEHRP.
- The management, coordination, and implementation of NEHRP.

The ACEHR is composed of leading earthquake professionals representing a balance of research and practitioner expertise, of regional, state, and local interests, and the relevant elements of the private sector. The ACEHR met for the first time in May 2007, at which members were updated on activities at the four NEHRP agencies and discussed their views on trends and developments with potential NEHRP impacts. The ACEHR met for the second time in October 2007.

Program Coordination Working Group

The Program Coordination Working Group (PCWG) is composed of working/level program managers from each NEHRP agency. The PCWG, chaired by NIST, meets monthly to coordinate agency activities, review reporting and planning documents, discuss issues and joint opportunities, and exchange relevant information. NIST maintains the NEHRP Secretariat to support PCWG activities.

The PCWG supports the efforts of the ICC in the preparation of the NEHRP annual reports and the revised Strategic Plan. In May 2007, the PCWG briefed staff members of the House Committee on Science and Technology and of the Senate Appropriations Committee on NEHRP activities. Through the PCWG, the NEHRP agencies jointly sponsored a research and implementation needs workshop for improved seismic safety for existing buildings. The agencies also began planning for jointly sponsored FY 2008 workshops on earthquake impact scenarios, performance/based seismic design, and post/earthquake information management. A primary goal of the workshops is to engage both the earthquake research and professional communities in defining needs in key areas that should receive increased Program emphasis.

1.4. Program Highlights

Federal Emergency Management Agency

FEMA continued to develop tools, and associated training materials, for design professionals, contractors, and decision/makers to address a number of existing challenges in seismic safety. These efforts included addressing the rehabilitation of existing buildings, updating the Homebuilders' Guide to Earthquake Resistant Design and Construction (FEMA 232), which references the seismic provisions of the International Residential Code, and developing the Design Guidefor Improving Hospital Safety in Earthquakes, Floods, and High Winds (FEMA 577), which provides state/ of/the/art knowledge on the vulnerabilities faced by hospitals. As a result of FEMA's success in influencing the Nation's model building codes, states and localities are increasingly adopting and enforcing earthquake/resistant building codes. In FY 2007, there was a 30 percent increase in the number of jurisdictions in high/seismic regions that adopted a disaster/resistant building code. All 50 states have now adopted a national model building code, either in whole or in part.

National Institute of Standards and Technology

Supported by initial funding under the President's American Competitiveness Initiative, NIST reestablished its earthquake engineering research program. As FY 2007 came to a close, NIST was recruiting new research staff and awarding a multiyear, multitask research contract to a joint venture of the Applied Technology Council and the Consortium of Universities for Research in Earthquake Engineering.

The NEHRP Secretariat exerted substantial effort in supporting ICC, ACEHR, and PCWG activities during FY 2007. The Secretariat also raised the profile of NEHRP by updating the NEHRP Web site (www.nehrp.gov), expanding the NEHRP listserv, and publishing a monthly electronic newsletter entitled SeismicWaves. Ten editions of this publication have been posted to date, covering subjects ranging from a homeowners' guide for earthquake safety to pipeline survivability in earthquakes.

The Secretariat also conducted a number of outreach activities, participating in meetings with regional earthquake consortia, earthquake professional groups, and government organizations. The Secretariat also participated in a Congressional Natural Hazards Caucus seminar on earthquake risk in the New Madrid region.

National Science Foundation

At the end of FY 2007, three NSF/supported research centers—the Multidisciplinary Center for Earthquake Engineering Research, the Mid/America Earthquake Center, and the Pacific Earthquake Engineering Research Center—completed 10 years of NSF support. All three centers will continue as centers through various combinations of university, state, and private sector support, and with other federal funding. Through NSF support, these centers have made major contributions to the development of performance/based seismic design; improved fundamental understanding of seismic performance of structures ranging from buildings, bridges, and acute care facilities to critical utility lifelines; and developed advanced technologies to improve earthquake mitigation and response.

Through four annual program solicitations and the Small Grants for Exploratory Research program, as of September 2007, the NSF has funded over 40 research projects to use the George E. Brown, Jr. Network for Earthquake Engineering Simulation (NEES) facilities to study soil foundation and structure interaction; seismic performance of foundations, lifelines, and reinforced concrete, masonry, wood, and composite structures; behavior of steel frames that include innovative bracing schemes; seismic design of nonstructural systems; seismic risk mitigation of ports and harbors; seismic performance of bridge systems with conventional and innovative materials; and tsunami generation and impacts on the built environment.

As a part of NEES operations, the NEES Consortium, Inc. (NEESinc) continues to push the forefront of experimental testing with development and improvement of the experimental infrastructure through new telepresence (joint research through the Internet) and hybrid simulation tools. Hybrid simulation is an experimental technique where a substructure is experimentally tested at full/scale, or near full/scale, while the remainder of the structure is simulated by computers. To date, NEESinc has developed two hybrid simulation software tools: SimCor 2.5 and OpenFresco 2.0.

U.S. Geological Survey

The USGS continued its deployment of the Advanced National Seismic System (ANSS) that integrates, modernizes, and expands seismic monitoring, data analysis, and earthquake notification in the United States. By the end of FY 2007, ANSS had grown to 784 stations, which is just over 10 percent of the planned 7,100/station network.

The number of stations in the Global Seismographic Network, which is jointly supported by the USGS and the NSF, rose to 147. Telemetry upgrades improved communications capabilities at 36 of 39 planned sites. The USGS began issuing rapid estimates of population exposure to surface shaking through its Prompt Assessment of Global Earthquakes for Response system to help emergency responders and aid agencies assess the impact of an earthquake disaster within a few tens of minutes after a large earthquake.

New national seismic hazard maps were provided to the Building Seismic Safety Council for its consideration in developing the 2009 version of the NEHRP Recommended Provisions or Seismic Regulations or New Buildings and 0ther Structures. These maps will be considered for inclusion in the 2012 version of the International Building Code. For California, a new seismic forecast model was developed by the Working Group on California Earthquake Probabilities, a collaboration among the Southern California Earthquake Center (SCEC), the California Geological Survey, and the USGS. The USGS also released new urban seismic hazard maps for Seattle, the first probabilistic seismic hazard maps based on three/dimensional simulations of earthquake ground motions.

A new Multi/Hazard Demonstration Project in Southern California began in FY 2007. This project includes a systematic investigation of the earthquake history of the southern San Andreas Fault in partnership with SCEC that will contribute to an urban hazard assessment for the Los Angeles region. The project also developed a realistic scenario for a major earthquake on the southern San Andreas Fault, which will be the basis for the 2008 Golden Guardian training exercise.

1.5. Structure of this Report

The new NEHRP Strategic Plan that is in the final stages of development serves as the framework for this report. The Strategic Plan defines three Program goals and standards for the operation of NEHRP facilities, all of which closely track the activities defined by Congress for the Program in the 2004 reauthorization. Objectives within each goal define activities, expected results, and outcomes for the 5/year strategic planning period (FY 2008–2012). In this report, NEHRP accomplishments for FY 2007 are described for each of the Strategic Plan objectives and facility operations, providing a baseline for NEHRP as it enters the implementation phase of the FY 2008– 2012 strategic planning cycle. Future annual reports will follow this reporting framework, providing a straightforward and simple means of tracking and evaluating NEHRP yearly performance.

2. PROGRAM BUDGETS FOR FY 2008 AND FY 2009

2.1. Introduction

The fiscal year (FY) 2008 and proposed FY 2009 Program budgets are presented below in terms of funds directed toward or requested for National Earthquake Hazards Reduction Program (NEHRP) goals, as defined in the new Strategic Plan. Each goal is associated with a NEHRP "Program Activity," defined in Public Law 108/360, Section 103(2). Table 2.1 shows these relationships. In addition, this legislation authorized the development, operation, and maintenance of certain NEHRP facilities: the Advanced National Seismic System (ANSS), the George E. Brown, Jr. Network for Earthquake Engineering Simulation (NEES), and the Global Seismographic Network (GSN). Table 2.1 shows the relationships between congressionally defined activities and the major elements of the new Strategic Plan.

Table 2.1. Relationships of NEHRP Strategic Goals to Statutory Program Activities

NEHRP Strategic Goals	NEHRP Program Activity (as defined by Congress in P.L. 108/360)
Goal A: Improve understanding of earthquake processes and impacts.	Improve the understanding of earthquakes and their effects on communities, buildings, structures, and lifelines, through interdisciplinary research that involves engineering, natural sciences, and social, economic, and decision sciences.
Goal B: Develop cost/effective measures to reduce earthquake impacts on individuals, the built environment, and society at large.	Develop effective measures for earthquake hazards reduction.
Goal C: Improve the earthquake resilience of communities nationwide.	Promote the adoption of earthquake hazards reduction measures by federal, state, and local governments.
Develop, operate, and maintain NEHRP facilities.	Develop, operate, and maintain ANSS, NEES, and the GSN.

2.2. NEHRP FY 2008 Enacted Budgets Listed by Program Goal

Table 2.2 lists the enacted FY 2008 NEHRP agency budgets by Program goal.

Table 2.2. NEHRP FY 2008 Enacted Budgets

Program Goal	Funds Allocated to Goal ($M)1				
	FEMA2	NIST	NSF	USGS	Total
Goal A: Improve understanding of earthquake processes and impacts.	0.1	0.2	30.0	10.8	41.1
Goal B: Develop cost-effective measures to reduce earthquake impacts on individuals, the built environment, and society at large.	3.1	1.0		30.2	34.3
Goal C: Improve the earthquake resilience of communities nationwide.	2.9	0.5		3.9	7.3
Develop, operate, and maintain NEHRP facilities:					
ANSS—USGS				8.8	8.8
NEES—NSF			22.1		22.1
GSN—NSF and USGS			3.5	4.4	7.9
Total:	6.1	1.7	55.6	58.1	121.5

Notes on Table 2.2:
[1] Budgets are rounded to nearest $0.1M.

[2] The FEMA FY 2008 budget is an estimated allocation from the Department of Homeland Security (DHS) appropriation, which covers Program activities but excludes Salaries & Expenses (S&E) and State Grants administered by the FEMA Preparedness Directorate.

2.3. NEHRP FY 2009 Requested Budgets Listed by Program Goal

Table 2.3 lists the President's requested FY 2009 NEHRP agency budgets by Program goal.

Table 2.3. NEHRP FY 2009 Requested Budgets

Program Goal	Funds Allocated to Goal ($M)1				
	FEMA2	NIST	NSF	USGS	Total
Goal A: Improve understanding of earthquake processes and impacts.	0.1	0.8	30.0	9.9	40.8
Goal B: Develop cost-effective measures to reduce earthquake impacts on individuals, the built environment, and society at large.	3.1	4.5		27.5	35.1
Goal C: Improve the earthquake resilience of communities nationwide.	5.4	1.1		3.7	10.2
Develop, operate, and maintain NEHRP facilities:					
ANSS—USGS				8.0	8.0
NEES—NSF			22.9		22.9
GSN—NSF and USGS			3.5	4.0	7.5
Total:	8.6	6.4	56.4	53.1	124.5

3. EHRP FY 2007 ACTIVITIES AND RESULTS

This chapter is organized by the goals and objectives in the National Earthquake Hazards Reduction Program (NEHRP) Strategic Plan for fiscal years (FY) 2008–2012. The goals are directly related to the NEHRP activities defined in Public Law 108/360, Section 103(2). The activities and accomplishments described below are not listed under individual NEHRP agencies, as done in past reports, but under NEHRP objectives as combined efforts. This demonstrates how the NEHRP agencies are working toward the goals and objectives of the NEHRP Strategic Plan.

3.1. Strategic Goal A: Improve Understanding of Earthquake Processes and Impacts

This strategic goal is directly related to the congressionally defined NEHRP program activity "Improve understanding of earthquakes and their effects on communities, buildings, structures, and lifelines through interdisciplinary research that involves engineering, natural sciences, and social, economic, and decision sciences." The research supported and undertaken under Goal A provides a strong foundation for the development and implementation of practical earthquake risk reduction measures pursued under Goals B and C. Program accomplishments for FY 2007 are presented under the four objectives established for Goal A.

Objective 1: Advance Understanding of Earthquake Phenomena and Generation Processes Physics-Based Models of Earthquake Sources in Southern California

The Southern California Earthquake Center (SCEC), led by the University of Southern California, is a university consortium jointly supported by the National Science Foundation (NSF) and the U.S.

Geological Survey (USGS). Scientists at SCEC have developed consensus models of the structure of the Earth's crust beneath southern California. These sophisticated computer/based models include a new Community Fault Model that represents more than 160 active faults in southern California. This model, in turn, has been extended to produce a realistic model of the geology of the crustal blocks that lie between these structures. Combined, these models can be used in studies of earthquake potential and impacts. SCEC also has developed regional earthquake likelihood models for earthquake predictability experiments. Earthquake ground motion simulations have been used to improve attenuation relationships and create realistic earthquake scenarios. New types of laboratory experiments at SCEC institutions have helped to elucidate the frictional resistance during earthquakes. These results have been combined with field studies on exhumed faults to develop better models of dynamic rupture. Results of these and similar experiments will eventually be incorporated into the regional models and may provide a useful tool for earthquake forecasting.

Year and magnitude of latest large San Andreas earthquakes. Illustration courtesy of Earthquake Country Alliance.

Precise Mapping of Major Earthquake Fault Systems

In the San Francisco Bay Area, the NSF and the USGS cooperated with other partners in supporting the acquisition of Light Detection and Ranging (LiDAR) surveys of the major earthquake faults. This airborne technology is a cost/effective and precise method of mapping the ground surface with a resolution of vertical and horizontal position to within several centimeters, revealing the surface expression of geologic faults. In the spring of 2007, the NSF EarthScope program (a related non-NEHRP activity) acquired 1,400 square kilometers of high/resolution LiDAR imagery, typically using 1/kilometer/ wide swaths, along the major earthquake faults, including the San Andreas, Calaveras, San Gregorio, Rogers Creek, and Maacama Faults. The USGS partnered with the San Francisco Public Utilities Commission, Pacific Gas & Electric Company (PG&E), and the City of Berkeley to extend the survey to the urban sections of the Hayward Fault. The bare/earth models derived by stripping away overlying vegetation should lead to improved mapping of fault strand locations and landslides in the foothills of the East Bay.

Similarly, in the Portland, Oregon, metropolitan area, approximately 4,200 square kilometers of high/resolution LiDAR imagery was collected during the winter of 2006–2007. The USGS partnered with the Oregon Department of Geology and Mineral Industries, Portland Metro, and the Oregon Department of Forestry to establish the Oregon LiDAR Consortium and fund this data acquisition. The data collected, when merged with existing LiDAR data, will provide complete coverage of the Portland urban growth area as well as much of the adjacent region. Landslide mapping and searching for evidence of recent geologic faulting or surface deformation constitute two of the high/priority uses for the new data.

Tracking Tremors in the Pacific Northwest

One of the most interesting new discoveries in the earth sciences has been the observation of slow aseismic slip and nonvolcanic tremor, particularly in the Pacific Northwest and other subduction zone settings such as Japan. The tremendous growth in the types and number of monitoring instruments in the region continues to drive these discoveries. "Slow slip" refers to the relative motion across a fault that occurs over many seconds to months, without radiating seismic waves, and is measurable using geodetic methods employing global positioning satellites (GPS) and sensitive strain meters. "Tremor" refers to seismic waves that radiate from sources at very low amplitudes for long durations, detectable most easily on sensors tuned to record rapid, high/ frequency waves. Slow slip and tremor generation represent modes of fault motion that differ significantly from that which occurs during earthquakes, and thus represent new elements to the energy budget that must be balanced to assess the probability of major earthquakes. Some observations suggest the two phenomena are highly coupled, and they repeat with a regularity rarely observed in geologic processes. Results to date suggest that slow slip and tremor reflect conditions on the boundaries of fault areas where earthquake slip may occur, so that improved observations may clarify those faults most likely to host future damaging earthquakes. NEHRP's sustained program of seismic monitoring in the region promises to provide new observations of and insights into these phenomena.

USGS External Grants and Cooperative Agreements

In 2007, the USGS funded over 100 grants and cooperative agreements to support targeted research in earthquake hazards, physical processes, and effects. This assistance adds a significant range of expertise to the USGS Earthquake Hazards Program and leverages support from other federal and state agencies, universities, and the private sector. SCEC, which is supported jointly with NSF, was awarded a 5/year cooperative agreement in FY 2007, thereby extending this successful partnership. Lists of USGS grants issued and reports that describe research results are available at www.earthquake.usgs.gov/research/external/.

Objective 2: Advance Understanding of Earthquake Effects on the Built Environment

Liquefaction (Ground Failure) Mitigation in Earthquakes
When loose, unconsolidated soils are shaken in earthquakes, they can lose bearing strength and flow or liquefy, thus damaging structures built upon them in the flow path. Research sponsored by the NSF at Northwestern University has been investigating the use of entrapped air or gas bubbles to mitigate liquefaction in loose granular soil deposits. Results to date show that small amounts of entrapped air can be introduced in saturated granular soils, and remain for long periods, even under very large flow gradients in any direction. As a result, this technique of induced partial saturation has the potential for long/term liquefaction mitigation. The construction of a testing apparatus is complete and initial testing has shown significant reduction in liquefaction with very small decreases in degree of saturation. Work continues to quantify the effects of entrapped air and evaluate field methods of inducing partial saturation.

In a related study, USGS researchers have improved understanding of liquefaction processes that can cause severe damage to structures during an earthquake. By analyzing borehole pore pressure data, they found that long/period seismic waves confined to near the Earth's surface, in addition to seismic shear waves traveling through the interior of the Earth, can produce excess pore pressure and liquefaction. Thus, conventional techniques for predicting liquefaction, based only on the interior waves, need to be amended to account for the importance of long/period surface waves.

Because surface waves carry higher amplitudes to greater distances, this has important implications for estimating the maximum distance from an earthquake where liquefaction can occur.

Performance-Based Seismic Design (PBSD) Research
Through its performance/based earthquake engineering methodology and tools, the Pacific Earthquake Engineering Research (PEER) Center has impacted both the development of future performance/based design standards and refinements to existing codes and standards for the building, transportation, and utilities industries. This is evident through its impact on projects funded by the Federal Emergency Management Agency (FEMA) to develop guidelines for PBSD. The PEER/developed software, Open System for Earthquake Engineering Simulation (OpenSees), provides a computational platform for advanced geotechnical and structural modeling that now has a large user base

(http://opensees.berkeley.edu). OpenSees has been adopted by the George E. Brown, Jr. Network for Earthquake Engineering Simulation (NEES) program.

Risk and Consequence Management for Communities
The Mid/America Earthquake (MAE) Center has developed methodologies and tools to address the complex impacts of earthquakes on vulnerable communities through a concept called "consequence– based risk management." Its approach is based on innovative and integrated application of its disaster impact assessment, mitigation, response, and recovery developments. The MAE Center has achieved major developments in engineering, social, and economic sciences, with new products coordinated into its MAEviz, an earthquake impact assessment and decision/making software package developed by the Center (http://mae.cee.uiuc.edu/software and tools/maeviz.html).

Hybrid Structure Simulation and Data Display and Sharing Tools
As a part of NEES operations, the NEES Consortium, Inc. (NEESinc) continued to push the forefront of experimental testing with development and improvement of the experimental infrastructure through new telepresence tools and hybrid simulation. New telepresence tools include three/dimensional viewers of experimental data developed by Rensselaer Polytechnic Institute and the University of California (UC)-Davis. Hybrid simulation is an experimental technique where a substructure is experimentally tested at full/scale, or near full/scale, while the remainder of the structure is simulated by computers. To date, NEESinc has developed two hybrid simulation software tools: SimCor 2.5, developed by the University of Illinois at Urbana/ Champaign, and OpenFresco 2.0, developed by the UC-Berkeley.

Objective 3: Advance Understanding of the Social, Psychological, and Economic Factors Linked to Implementing Risk Reduction and Mitigation Strategies in the Public and Private Sectors

Analysis of Post-Disaster Commodity Lifelines
The NSF has supported research relevant to earthquake response in the study, "Characteristics of the Supply Chains in the Aftermath of an Extreme Event: The Gulf Coast Experience." This multidisciplinary effort at the Rensselaer Polytechnic Institute and the University of Delaware's Disaster Research Center undertook a study to analyze the provision of supplies and aid to victims of the Gulf Coast after Hurricane Katrina. The findings from this collaborative team of

engineers and social scientists indicated that supply chains were inadequate. The primary factors associated with the failure to deliver needed supplies, materials, and expertise included the magnitude of the requirements; collapse of the communication infrastructure; understaffing and lack of training; lack of integration between federal and state logistic systems; inefficiencies in pre/positioning resources; lack of planning for handling and distribution of donations; procurement; limited asset visibility; network visibility; and social vulnerability. The findings appear to be applicable to the post/impact periods of other extreme events, such as earthquakes.

Improving Urban Search and Rescue Preparedness Using Simulation Technology

Researchers from the University of Michigan and the University of Delaware have collaborated on a study of search and rescue activity at collapsed structures and in situations involving fire following earthquakes. An attempt is being made to understand patterns of collapse in relationship to activities undertaken by occupants to improve their chances of survival. The goal is to create effective tools using virtual reality visualization to train urban search and rescue personnel. In addition, the knowledge that is produced will be used as a basis of public education and awareness programs to help untrained volunteer rescuers.

Measuring Cross-Community Disaster Preparedness and Resiliency

Researchers at the University of Louisville are developing metrics for assessing community/level emergency preparedness and resilience. The goal is to develop a Community Disaster Preparedness Index (DPI) that will allow for uniform, valid, and reliable comparisons across communities with regard to their preparedness for managing multiple hazards, including earthquakes. In addition, the DPI is being developed in conjunction with the development of a Community Disaster Resilience Index (DRI) that will allow for comparative assessment of the potential for communities to recover from extreme events, such as earthquakes. The DPI and DRI indices will provide an opportunity to make meaningful cross/comparisons among communities regarding their emergency preparedness, response capabilities, and potential for recovery.

The Effects on Natural Hazard Mitigation of New Urban Developments Compared to Conventional Low-Density Developments

Land/use regulation and urban design are key components of nonstructural hazard mitigation measures. Researchers in the Department of City and Regional Planning at the University of North Carolina at Chapel Hill are undertaking a comparative analysis of communities that include design and land use patterns consistent with "New Urbanism" and those with conventional, low/density land use patterns. The inherent vulnerability to natural hazards, including earthquakes, of New Urbanism communities as opposed to conventional communities is being assessed. The extent to which hazard mitigation practices are integrated into site designs for New Urbanism communities, as opposed to conventional communities, is also being investigated. The results can have a significant impact upon mitigating future earthquake losses through sound land/use planning and governance.

Objective 4: Improve Post-Earthquake Information Management Learning Rom Earthquakes Program

During the past 5 years, the Earthquake Engineering Research Institute (EERI), with NSF support, has sent teams of structural engineers, geotechnical engineers, and social scientists to the scene of significant earthquakes throughout the world to gather ephemeral and perishable data. Dates and locations of earthquakes studied were 2002 Molise, Italy; 2002 Denali, Alaska; 2003 Bam, Iran; 2003 Colima, Mexico; 2004 Niigata, Japan; 2004–2005 Sumatra and Indian Ocean (seven teams); 2005 Kashmir, Pakistan; 2005 Tarapaca, Chile; 2006 Hawaii; 2006 Indonesia; and 2007 Peru.

Over the past 5 years, more than 680 collaborators have investigated and documented a total of 27 earthquakes and published more than 40 articles, books, and research reports. Results are available at www.eeri.org/lfe.html. These investigations will provide the basic data for the Post/Earthquake Information Management System (PIMS) that will be developed by NEHRP in the future.

Planning or a PIMS

The American Lifelines Alliance held a forum on October 11–12, 2006, at the American Institute of Architects headquarters in Washington, D.C. The forum, attended by over 50 representatives of the lifelines community, focused on post/disaster data collection needs for critical infrastructure lifelines following earthquakes and other disasters. Participants worked on the process for developing a framework for improving the mechanisms for the collection, management, and archiving of data on the performance of the built environment in natural disasters within the United States. A report of the forum has been completed and a follow/on event is planned for FY 2008.

3.2. Strategic Goal B: Develop Cost-Effective Measures to Reduce Earthquake Impacts on Individuals, the Built Environment, and Society at Large

Work under Goal B builds on the research results of Goal A to develop practical and cost/effective methods and measures for earthquake risk assessment and mitigation. Goal B is directly linked to the congressionally defined NEHRP program activity "Develop effective measures, i.e. policies and practices, for earthquake hazards reduction."

Objective 5: Assess Earthquake Hazards for Research and Practical Application Updated National Seismic Hazard Maps

The USGS National Seismic Hazard Mapping Project provided new hazard maps to the Building Seismic Safety Council (BSSC) for its consideration in the 2009 version of the NEHRP Recommended Provisions or Seismic Regulations or New Buildings and 0ther Structures (NEHRP Recommended Provisions). These maps will be considered for inclusion in the 2012 version of the International Building Code (IBC). The maps and related hazard curves were developed using the best available science based on internal USGS studies and information from government agencies, academic institutions, and industry, including updated attenuation equations for how ground motions decrease with distance from faults. For the Western United States, these attenuation relations came from the PEER Center Next Generation Attenuation Project, funded by the NSF and other partners. For some measures of ground shaking, these new relations predict lower amounts of shaking for large earthquakes than were previously estimated. For California, the Working Group on California Earthquake Probabilities, based on a collaboration between SCEC, the California Geological Survey (CGS), and the USGS, developed a new seismic forecast model.

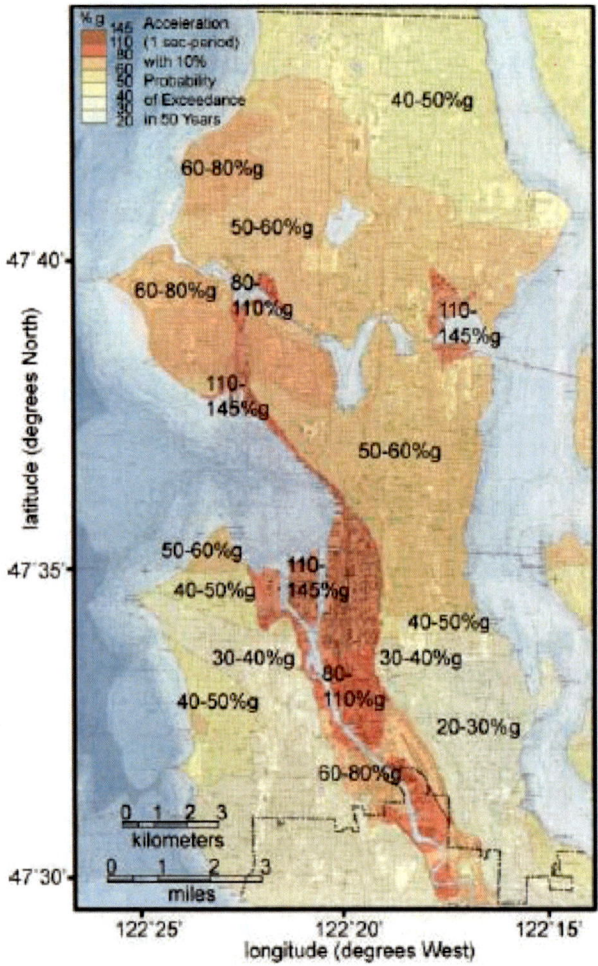

Map showing the expected ground shaking due to an earthquake on the Seattle Fault. The map is based on a three-dimensional model of the regional geology and fault geometry. The lateral force of the ground shaking is given as a percentage of building weight. Image courtesy of USGS.

Seismic Hazard Maps for the Seattle Urban Region

The USGS released new urban seismic hazard maps for Seattle in July 2007. These maps depict the ground shaking at three different levels of probability, appropriate for assessing the hazard to medium/sized buildings and some bridges. They are the first probabilistic seismic hazard maps based on three/dimensional simulations of earthquake ground motions. Over 500 computer simulations of earthquakes on the Seattle Fault, the Cascadia Subduction Zone, the South Whidbey Island Fault, and background shallow and deep source zones were used. The detailed maps incorporate the amplification of ground shaking by the deep structure of the Seattle basin and the shallow layers of artificial fill, and include the buildup of ground shaking caused when earthquake rupture is directed upwards along a dipping fault. The maps were presented to the Seattle City Council and the Washington State Department of Transportation (WSDOT). Many uses have already been proposed for the maps, including the prioritization of seismic retrofit for unreinforced masonry buildings and the preliminary design of a major bridge project.

New ShakeMaps for the 1868 Hayward Fault Earthquake

New ShakeMaps that show the distribution and intensity of ground shaking have been developed for the large 1868 earthquake on the Hayward Fault in the East Bay region of northern California. The ShakeMaps are based on historical data acquired in cities in and around the Hayward Fault. The new data indicate that the 1868 earthquake was centered in the Hayward area and that its magnitude was larger than previously estimated ($\approx$M7.0). Because nearly 2 million people now live near the Hayward Fault, the new data clearly raise concerns. These maps will be used for preparedness exercises in 2008.

Multi-Hazard Demonstration Project in Southern California

As part of its Multi/Hazard Demonstration Project in southern California, the USGS will continue a systematic investigation of the earthquake history of the southern San Andreas Fault in partnership with SCEC. This analysis will contribute to an urban hazard assessment for the Los Angeles region to be completed in 2009. The goal of the broader Multi/Hazard Demonstration Project is to link research results and data with information dissemination to provide an integrated approach to hazards research, warning, and mitigation. This multiyear effort focuses on the eight counties of southern California, where catastrophic losses from natural hazards such as earthquakes, tsunamis, fires, landslides, and floods exceed $3 billion per year. Partners include state, county, city, and public lands government agencies, public and private utilities, industry, academic researchers, FEMA, the National Oceanic and Atmospheric Administration (NOAA), the U.S. Forest Service, the Bureau of Land Management, and local emergency response agencies.

Objective 6: Develop Advanced Loss Estimation and Risk Assessment Tools Further Development of hazards U.S.-Multihazard (HAZUS-MH)

In 1997, FEMA produced through a contract with the National Institute of Building Sciences (NIBS), the prototype HAZUS97 software, a nationally applicable, standardized, computer/based disaster planning and analysis tool. It can be used to estimate sizes and locations of possible threats; calculate resulting damage and disruption; apply supporting data from varied sources; and link with other emergency management and planning tools before, during, and after disasters. Since then, a multihazard version called HAZUS-MH, has helped communities identify and plan for earthquakes by giving them free access to specialized databases and geographic information system (GIS)-based tools.

The HAZUS-MH Earthquake Model provides estimates of damage and loss to buildings, essential facilities, transportation lifelines, utility lifelines, and population based on scenario or probabilistic earthquakes. The advanced HAZUS-MH MR2 Earthquake Model includes custom building types and permits importing USGS ShakeMaps and optimized software for faster performance. The new HAZUS-MH MR3 Earthquake Model includes adjustable population distribution parameters in the casualty module. The Comprehensive Data Management System allows users to update and manage statewide data sets, which are now used to support analysis in HAZUS-MH.

In FY 2007, FEMA Headquarters and the FEMA Regions continued to actively promote and support state and local use of HAZUS-MH. It was used to support disaster scenarios for catastrophic planning exercises in the New Madrid region of the central Mississippi Valley and elsewhere, in preparedness training sessions and workshops, and to support the establishment and operation of HAZUS users groups.

Risk Assessment for the Northeast United States

The Northeast States Emergency Consortium (NESEC) operates the NESEC Emergency Management Risk Assessment Center (NEMRAC). NEMRAC's priority is to provide GIS and HAZUS-MH support to jurisdictions that do not have the resources and staff to develop HAZUSMH or GIS products in/house. NESEC provides links and information on obtaining GIS software and technical manuals on its Web site, and has recently moved its HAZUS-MH/GIS services request system online at www.nesec.org/resources/. NESEC is developing an online Hazard/ Resistant Building Code Database that will allow the public to determine if their community has appropriate building code regulations for earthquakes and other hazards. The Hazard/Resistant Building Code Database will be made available to the public once building code content and usage data are obtained.

Objective 7: Develop Tools to Improve the Seismic Performance of Buildings and Other Structures

Building Standards and Codes: NEHRP Recommended Provisions

The NEHRP Recommended Provisions, comprising the Provisions (Part 1) and Commentary (Part 2), is the primary resource document for the Nation's standards and model codes for new buildings. Current project work is focused on completing the 5/year update process that will result in the 2009 edition of the NEHRP Recommended Provisions. The update features several significant changes that have been recommended or endorsed by the practitioner community. The model building codes have adopted by reference significant portions of the seismic chapters of the American Society of Civil Engineers (ASCE) standard, Minimum DesignLoads for Buildings and 0ther Structures (ASCE 7), which relies heavily on the NEHRP Recommended Provisions for much of this material. As a result, the NEHRP Recommended Provisions now serves as a research/to/practice resource document. As part of the revision process, FEMA is completely rewriting the Commentary so that it can better serve as a training and educational product. This collaborative project includes the USGS for developing new seismic hazard and design maps; the National Institute of Standards and Technology for reviewing drafts and performing applied research; earthquake professionals for providing engineering knowledge based on practical experience; and researchers supported by the NSF for providing basic research input.

PBSD Development

The PBSD project is a multiyear effort to develop PBSD guidelines for new and existing buildings. The project goal is to develop practical assessment and design criteria that enable building owners and regulators to select desired performance levels for new construction or for upgrades of existing buildings that differ from the current building code/based life/safety level.

The development work is being done under contract with the Applied Technology Council (ATC), an organization that develops engineering resources for use in mitigating the effects of natural and other hazards on the built environment.

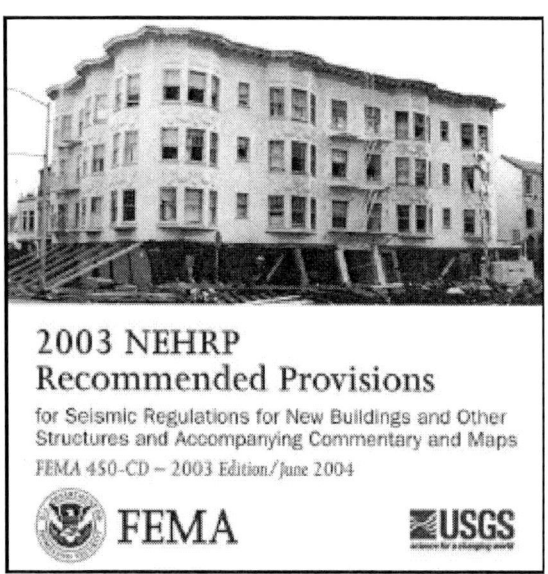

FEMA 450 CD cover courtesy of FEMA.

FEMA is now in the second year of a 5/year effort to develop the PBSD Performance Assessment Calculation Tool (PACT) and associated guidance. PACT will be used to evaluate the performance of new and existing structures by applying standard methods of structural analysis, coupled with structural reliability and loss estimation methods. As part of this project, FEMA has published a new document that provides different methodologies on how to consistently test the performance of building components (FEMA 461, Interim Protocolsfor Determining the Seismic Performance Characteristics of Structural and Nonstructural Components). This publication was developed with the three NSF earthquake engineering research centers. In the final phase of the project, PACT will be used as the basis for developing the PBSD guidelines. Additional information on the PBSD project may be found at www.atcouncil.org/atc/58.shtml.

Seismic Rehabilitation of Existing Buildings Workshop

In September 2007, more than 90 earthquake professionals participated in a NEHRP-funded workshop in San Francisco: Meeting the Challenges of Existing Buildings. The purpose of the workshop was to identify key issues and potential strategies that the NEHRP agencies can use to determine how best to proceed with developing new guidance for mitigating earthquake risk to existing buildings. A key to the success of the workshop was the involvement and participation of the NEHRP agencies, representatives of NEESinc, academic researchers, and other earthquake professionals involved in research and practice. Practitioners in engineering and other earthquake disciplines had the opportunity to share information and views concerning critical research priorities that require use of the NEES experimental facilities. The workshop was conducted as a joint effort by the ATC and EERI.

Homebuilders' Guide to Earthquake Resistant Design and Construction
FEMA continues to promote the use of the recently updated FEMA 232, Homebuilders' Guide to Earthquake Resistant Design and Construction. FEMA 232 includes the latest changes to the International Residential Code (IRC) and the results of the Caltech Wood Frame Buildings Project conducted by the Consortium of Universities for Research in Earthquake Engineering. FEMA 232 also presents a series of "above/code" recommendations that have been shown to improve home performance in earthquakes and increase chances of post/earthquake habitability. FEMA recently sent the International Code Council (ICC), a partner on the publication, 5,000 copies to distribute free of charge to their members.

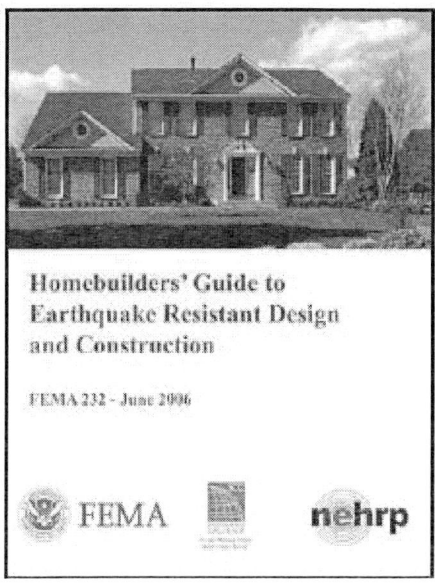

FEMA 232 cover courtesy of FEMA.

Design Manual for Architects
EERI published FEMA 454, Designing for Earthquakes: A Manual for Architects. FEMA 454 explains the principles of seismic design for those without a technical background in engineering and seismology, practicing architects, architectural students, and faculty in architectural schools that teach structures and seismic design.

Objective 8: Develop Tools to Improve the Seismic Performance of Critical Infrastructure

Improving the Seismic Performance of Infrastructure and Critical Facilities
In FY 2007, the Multidisciplinary Center for Earthquake Engineering Research (MCEER) developed the "Rehabilitation Decision Analysis Toolbox" for acute care facilities. The system is designed for use by hospitals as a decision/making tool on capital improvements needed to most effectively make them more hazard/resistant. The Center has also developed advanced systems analysis tools to predict the seismic resilience of power networks and water systems, and new materials and technologies for the seismic retrofit of pipelines, electrical transformers, buildings, and nonstructural building components.

Pipeline Safety
Improved seismic design of underground utility pipelines is needed to keep future earthquakes from becoming the two/stage disasters seen in the 1906 San Francisco and 1995 Kobe (Japan) earthquakes, where damage from shaking was followed by damage from fires that are ignited, fueled, or allowed to grow by ruptured gas or water lines. NSF/supported research using the NEES lifeline experimental facility at Cornell University has confirmed that the use of high/density polyethylene (HDPE) pipelines, which are being used by industry in a growing number of sizes and settings, in earthquake/prone areas would help prevent earthquake/induced pipeline ruptures and their potentially catastrophic consequences. HDPE, a type of plastic, has demonstrated that it can stretch and deform without breaking when strained by extreme forces. This research is producing findings that can lead to better, more earthquake/resistant pipelines as the findings are put into practice.

Annual Report of the National Earthquake Hazards Reduction Program 57

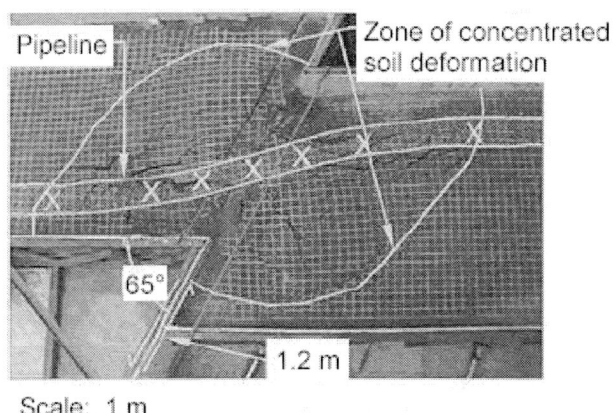

Overhead view of a large-scale fault-rupture test on a 400-mm-diameter HDPE pipeline at the Cornell University NEES laboratory; X's mark locations of sensors. Image courtesy of Cornell University.

Multispan Bridge Safety

With NSF support, three NEES shake tables at the University of Nevada, Reno (UNR) are being used to test, for the first time, the seismic performance of entire four-span bridges along with the performance of individual bridge components. The project is testing three models, at about one- fourth the size of real bridges, which are made with different materials and design details.

The first bridge model, a conventional reinforced-concrete structure about 33.53 meters (110 feet) long, was tested at UNR in February 2007. Over the next 2 years, researchers plan to test two more bridge models that incorporate innovative seismic-resistant features. Reinforced-concrete piers will be strengthened with outer layers made from fiber-reinforced polymer composites; columns will be made more resilient with built-in isolators designed to absorb and damp down seismic forces; and column-footing connections will be toughened with new "superelastic" nickel titanium rods and flexible concrete.

3.3. Strategic Goal C: Improve the EarthquakeResilience of Communities Nationwide

Work under Goal C transfers research results (Goal A) and development of methodologies (Goal B) to the support and promotion of practical implementation of earthquake risk mitigation measures. This goal is directly related to the congressionally defined NEHRP program activity "Promote the adoption of earthquake hazard reduction measures by federal, state, and local governments, national standards and model code organizations, architects and engineers, building owners, and others with a role in planning and constructing buildings, structures, and lifelines."

Objective 9: Improve the Accuracy, Timeliness, and Content of EarthquakeInformation Products

The accomplishments described under this objective are results and products based on the regional and national seismic networks and data centers of the Advanced National Seismic System (ANSS). ANSS is described more fully in Section 3.4.

ShakeCast Upgraded. User Base Expanded

Immediately following an earthquake, the USGS prepares ShakeMaps showing the distribution and severity of strong ground shaking in the vicinity of the earthquake epicenter. These maps are posted on various Web sites for user access. The ShakeCast system, which pushes ShakeMaps results out to users, has been significantly updated. The system delivers to user sites estimates of ground motion based on ShakeMap, as well as the underlying empirical data, in formats compatible with user requirements. The California Department of Transportation is probably the most advanced user of ShakeCast. Several other entities are also in various stages of installing capabilities to receive and display ShakeCast, including the U.S. Department of Veterans Affairs (for its medical center facilities), Sempra Energy, the Department of Water and Power with the City of Los Angeles, and the Los Angeles Unified School District.

This schematic diagram shows how ShakeCast combines ShakeMap input data with pre-defined structure fragilities to produce, display, and broadcast damage estimates. Image courtesy of USGS.

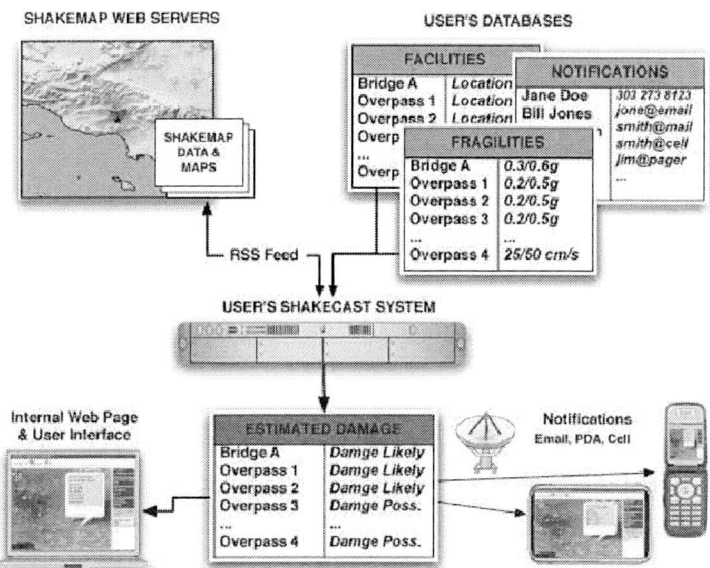

This schematic diagram shows how ShakeCast combines ShakeMap input data with pre-defined structure fragilities to produce, display, and broadcast damage estimates. Image courtesy of USGS.

BetaPAGER Development Completed

The Prompt Assessment of Global Earthquakes for Response (PAGER) is a new product of ANSS that helps emergency responders quickly gauge the impact of an earthquake disaster. PAGER provides, within a few tens of minutes after a large earthquake, estimates of population exposure to graded levels of surface shaking; these can be interpreted to indicate likely damage, depending on the earthquake resistance of the built environment at the earthquake site. PAGER development for population exposure was completed in 2007 and PAGER results are now being distributed by text message, e-mail, and Web pages to the public via betaPAGER. Current efforts are to translate shaking data and models into loss estimates.

Earthquake Alert Display Software Upgraded

Upgrades were completed to the California Integrated Seismic Network (CISN) Display software tool, including a "kiosk mode." The kiosk mode makes it possible to deploy the CISN Display as part of museum exhibits. CISN Display also now provides NOAA tsunami messages. Several hundred users, including seven television stations in the Los Angles area and Indian tribes in Washington State, are operating the software.

Earthquake Early Warning Technical Tests under Way

The USGS has funded a 3/year project in California to test earthquake early warning (EEW) algorithms. The project includes testing of algorithms using data from CISN stations at UCBerkeley, Caltech, and the Swiss Federal Institute of Technology in Zurich, Switzerland; research on EEW for quakes with large ruptures at Caltech; and comparison of timelines and results from algorithms at SCEC.

These algorithms use real/time seismic data to rapidly estimate an earthquake's location and magnitude, and predict the spatial distribution of shaking. EEW systems deployed in Japan and Mexico broadcast this information to some users before damaging waves arrive, providing precious seconds to prepare and mitigate loss. Tests under way will measure the precision of existing EEW algorithms, estimate the warning times possible for a deployed EEW system on the U.S. West Coast, and assess the improvements in infrastructure and processing necessary to construct an operational system.

Notification and Analysis of Utah Mine Collapse

On August 6, 2007, a shallow seismic event with a magnitude of 3.9 was recorded on seismic stations operated by University of Utah Seismic Stations (UUSS) and the USGS. The data from those stations were combined to locate the event in the vicinity of the Crandall Canyon Mine in the Wasatch Plateau coalfield, site of a tragic mine collapse that led to the deaths of several trapped mine workers. Seismologists at the UUSS, which operates a 180/station regional network as part of ANSS, and the National Earthquake Information Center (NEIC) quickly determined that the seismic recordings were consistent with a type of shock induced by underground mine collapse, rather than a natural earthquake. Because of the complexity of performing such analyses, seismologists at the USGS and the University of Utah then consulted with experts at the UC-Berkeley Seismological Laboratory. A more detailed analysis, which employed long/period seismic waves recorded on high/quality digital seismic stations, including ones recently deployed as part of the NSF's EarthScope program, confirmed the initial finding. Throughout the tragic unfolding of events at the Crandall Canyon Mine, NEHRP scientists worked closely with their counterparts at the University of Utah and elsewhere to ensure that officials, the media, and the public were provided with the best available scientific characterization of the seismic signals generated in the vicinity of the mine.

Annual Report of the National Earthquake Hazards Reduction Program   61

Earthquake Notification in the Northeast United States
NESEC, in partnership with the Massachusetts Emergency Management Agency, maintains a multi/state Earthquake Hazard Notification System to inform state emergency management agencies when earthquakes occur. In FY 2007, the system alerted officials of a dozen seismic events occurring in the Northeast. NESEC also created a real/time hazard maps Web site (www.nesec.org/hazard maps.cfm) that provides emergency managers and the public with links to USGS sites, allowing them to monitor earthquakes and other hazards in real time.

Operationsfor Monitoring Non-Seismic (Geodetic)
Movement of the Earth's Crust
The USGS issued several 3/year cooperative agreements for continuous GPS, strain, and creep monitoring operations. To better understand the behavior of faults, these operators measure crustal deformation based on the relative movement of points on the Earth's surface related to ground tilt, ground strain, and fault slip (creep). Data from the studies are made available for use by the research community.

Objective 10: Develop Comprehensive Earthquake Scenarios and Risk Assessments

Scenario Earthquake for the Southern San Andreas Fault
NEHRP-supported research led an effort to construct a realistic scenario for a major earthquake on the southern San Andreas Fault, which last ruptured in the 17th century. The rupture scenario, completed in 2007, will be the basis for the 2008 Golden Guardian training exercise. The scenario earthquake, developed by a collaboration of USGS and university geologists through the SCEC partnership, provides a realistic example of a future large earthquake on the San Andreas Fault. The magnitude/ 7.8 earthquake source extends 305 kilometers from the Mexican border to north of Los Angeles, with slip distributed according to a model based upon slip rates and fault geometry specified by the Working Group on California Earthquake Probabilities.

Using the predicted fault displacements from the scenario earthquake and established methodologies to predict the shaking levels throughout southern California, the resulting scenario will allow stakeholders in the region to consider in detail the potential impact of a future "Big One" in California.

The scenario work has highlighted significant lifeline vulnerabilities at key transportation arteries that cross the fault, and NEHRP researchers have been in close communication with lifeline operators to discuss results and concerns. The scenario earthquake source and estimates of resulting ground motions have already been used as the basis for a large/scale joint civil/military exercise, Golden Phoenix 2007, and for tabletop disaster response exercises by the City of Los Angeles and several area/level mutual aid first responder groups. The scenario development is taking place in conjunction with a major preparedness campaign, Dare to Prepare, which includes hundreds of partner organizations in southern California.

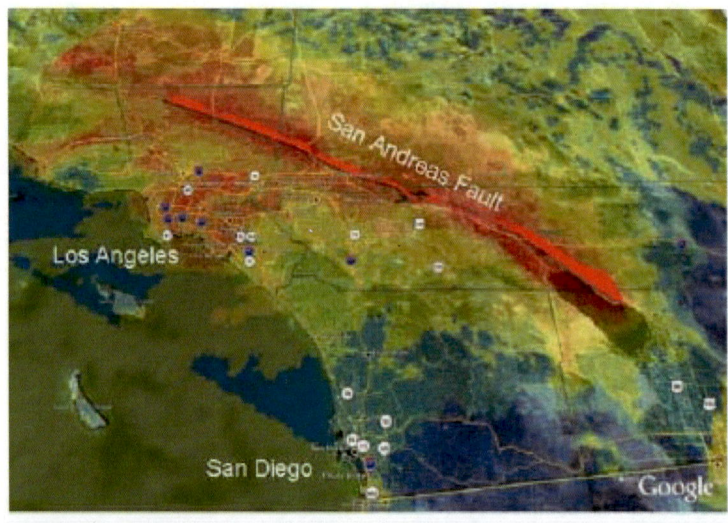

This scenario map shows the expected distribution and severity of ground shaking from a major earthquake on the San Andreas Fault in southern California. All of southern California would be affected by this earthquake. Image courtesy of the Southern California Earthquake Center at the University of Southern California.

California Catastrophic Disaster Readiness Planning

Planning for the California Catastrophic Earthquake Readiness Response Plan (CCERRP) began in July 2007. The first phase will produce a Concept of Operations that will clarify authorities between federal and state partners, integrate the doctrine and policy of the National Incident Management System and California's State Emergency Management System, and provide a statewide all/hazards framework for responding to a catastrophic event that exceeds California's considerable response capabilities. The second phase of the CCERRP will provide a more detailed Concept of Operations Plan (CONPLAN) that will be based on the need to respond to a potential recurrence of the 1906 Bay Area earthquake and resulting fires. This plan will also provide a schedule of materials needed by the state and will facilitate the delivery of timely assistance. On completion of the plans, FEMA and the California Office of Emergency Services will conduct exercises to validate them and will develop a CONPLAN for response to a southern California earthquake.

Earthquake Disaster Planning in the Central United States
Initiated in July 2006, the New Madrid Seismic Zone Catastrophic Planning Project will result in the first national plan for response to a New Madrid earthquake that integrates federal (national and regional), state, and local plans. The project involves unprecedented coordination, across government and among the agencies residing within each government level. FEMA's Disaster Operations Directorate is leading, with NEHRP support, a "local/up" approach to the effort, focusing on state and local priorities and integrating them with federal priorities to ensure all needs are met. The project's geographic scope requires consensus and close coordination among FEMA Headquarters (Disaster Operations, Disaster Assistance, Mitigation, and Preparedness), FEMA Regions IV, V, VI, and VII, and the States of Alabama, Arkansas, Kentucky, Illinois, Indiana, Mississippi, Missouri, and Tennessee. The national plan will serve as a baseline document for further development of federal, state, and local plans.

Disaster planning activities of the Central United States Earthquake Consortium (CUSEC) increased significantly as a result of the FEMA regional catastrophic planning effort. Activities included integration of components of risk analysis, building science, outreach, and community planning. Risk analysis projects included models for the region, seven state/specific runs, and numerous runs at the local level. Risk assessment also played a pivotal role in the 2007 Spills of National Significance exercise. CUSEC hosted planning meetings and was an active participant in the exercise.

Earthquake Disaster Planning in the Pacific Northwest

The Cascadia Region Earthquake Workgroup (CREW) is using the Cascadia Region Earthquake Subduction Zone Scenario to encourage the discussion of associated impacts and risk reduction measures. CREW is supporting the Pacific

Northwest Economic Region and the U.S. Department of Transportation in their use of the CREW scenario to drive the Blue Cascades 3 and Pacific Peril exercises, which focus on the Interstate 5 corridor and the Pacific Coast, respectively. CREW will consolidate results of these efforts and develop a comprehensive statement of impact as a companion document to the initial scenario. In cooperation with the University of Oregon and the Oregon Natural Hazards Workgroup, CREW also is developing and piloting a post/disaster recovery forum for the coastal communities.

Report to Congress on Federal Earthquake Response Plans
FEMA is developing a report to Congress on completed and in/progress federal earthquake response plans and planning at the national and regional levels. The report discusses the relationship between all/hazards plans and hazard/specific plans; the role of federal and regional offices in providing a strong link between state and federal personnel who staff the plans; and the need to develop several layers of plans to cover all the actions necessary to assist disaster victims. The report also discusses planning activities at the federal and regional level, training to ensure plans are understood and executable, and scenario/based exercises that reveal where improvement is needed within existing plans.

Objective 11: Support Development of Seismic Standards and Building Codes and Advocate Their Adoption and Enforcement

Support for Seismic Elements of New and Existing
Building Codes and Standards
FEMA supports a group of experts who work with NIBS to submit changes or improvements, developed under the NEHRP Recommended Provisions, as proposed changes to the Nation's model building codes including the International Codes series of the ICC. NIBS also monitors other proposed changes to the codes and provides testimony during the code change process to help ensure the earthquake/related code provisions are not degraded. Several proposed code changes were approved for inclusion in the 2007 amendments to the 2006 editions of the IBC and IRC. NIBS was also involved in the latest revision to the National Fire Protection Association Publication 255 of its Manufactured Housing Installation Committee, which was successful in adding seismic requirements to that standard.

Annual Report of the National Earthquake Hazards Reduction Program 65

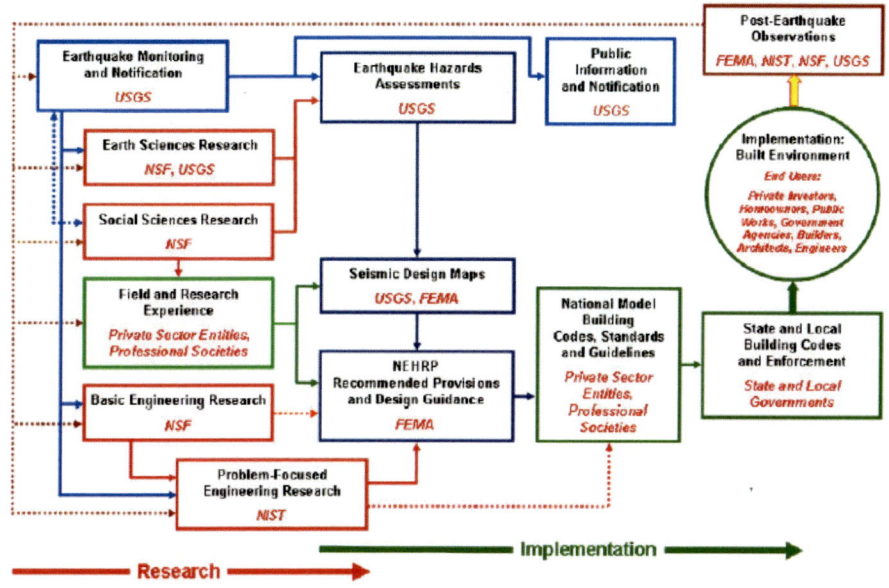

NEHRP Impact on the Built Environment. Image courtesy of NIST.

FEMA has provided testimony regarding the adoption of statewide building codes in South Carolina and Tennessee and worked with the ICC to develop code training materials, including an update to the seismic design edition of the popular ICC CodeMaster series. The NIBS BSSC Code Resource Support Committee is updating national seismic design maps for the model codes based on the 2007 USGS seismic hazard maps.

FEMA remains involved in the dissemination of ASCE 41/06, Seismic Rehabilitation of Existing Buildings. This new national performance/based consensus standard is based on FEMA 356, Pre- standard and Commentary for the Seismic Rehabilitation of Buildings. The completion of this standard culminated a 20/year NEHRP-wide effort led by FEMA. In addition to the new standard, a supplement to ASCE 41 will soon be released that incorporates the latest research results on the rehabilitation of reinforced concrete structures. Using the results of recent research and analysis, this supplement to the standard will significantly alter controversial elements of the reinforced concrete provisions in FEMA 356.

Objective 12: Promote the Implementation of Earthquake Resilient Measures in Professional Practice and in Private and Public Policies

NEHRP Recommended Provisions—Training and Instructional Material
NEHRP Recommended Provisions for New Buildings and Other Structures: Training and Instructional Materials, FEMA 451B-CD, was published in June 2007 in a compact disc format. FEMA 451B-CD contains a series of training slides and instructional material on the seismic design and construction of new buildings based on the NEHRP Recommended Provisions, FEMA 450, which serves as the basis for seismic requirements found in the Nation's model building codes and standards. It is a companion product to NEHRP Recommended Provisionsfor New Buildings and Other Structures: Design Examples, FEMA 451 CD, which contains a series of sample building designs for training purposes.

Earthquake Information Dissemination and Awareness
In FY 2007, the Natural Hazards Center (NHC) at the University of Colorado (www.colorado.edu/hazards/) distributed its quarterly newsletter, Unscheduled Events, to more than 16,000 subscribers. More than 3,000 individuals received its biweekly electronic newsletter. The NHC annual workshop each July brings together leading U.S. natural hazards researchers, policy makers, and practitioners. This is the major national forum for linking the producers of research with appropriate user communities. In FY 2007, nine other federal agencies contributed funds to support the NSF grant to the NHC, including FEMA and the USGS.

Technical Seminars on PBSD
EERI held the first in a series of technical seminars on "Performance-Based Earthquake Engineering for Structural and Geotechnical Engineers: Impact of Soil-Structure Interaction on Response of Structures" in Bellevue, Washington (91 attendees), Los Angeles (123 attendees), and San Francisco (127 attendees). Videos of the seminar presentations can be obtained, along with copies of each PowerPoint presentation, from the EERI Web site (www.eeri.org). Funding from FEMA supports EERI's 26 student chapters located at U.S. universities. The technical seminar videos are made available to the student chapters free of charge.

*Earthquake Policy Development and Coordination
in the Western United States*
In FY 2007, the Western States Seismic Policy Council (WSSPC) planned and held a joint conference with the ICC (September 30–October 3, 2007). Also

in FY 2007, WSSPC, along with other earthquake organizations, initiated planning for the National Earthquake Conference to be held in Seattle on April 22–26, 2008.

The joint publication Earthquake Hazards and Estimated Losses in the County of Hawaii, submitted by Hawaii State Civil Defense, the Hawaii State Earthquake Advisory Committee, and the Hawaii Coastal Zone Management Program, was the winner of the WSSPC 2007 Overall Award in Excellence. Six policy recommendations relating to tsunami notification, post-earthquake clearinghouses, seismic provisions in building codes, establishment of a Basin & Range Province Earthquake Working Group, and a post-earthquake information management system were proposed for adoption by the WSSPC membership at the 2007 annual conference.

State Earthquake Program Managers Meeting

In FY 2007, FEMA Headquarters, the FEMA Regions, and the regional consortia continued to support the implementation of effective practices and policies for earthquake loss reduction at the state and local levels. CUSEC, in collaboration with FEMA, NESEC, and the WSSPC, developed the agenda for the third annual State Earthquake Program Managers meeting that was hosted by CUSEC and the State of Tennessee. This activity continues to be the primary source for state-level program managers from across the United States and the territories to exchange information and share program success stories as well as lessons learned.

Earthquake Mitigation Training

Through the National Earthquake Technical Assistance Program (NETAP), FEMA supports the development of training curricula on earthquake mitigation topics and provides courses for state and local officials and businesses throughout the United States. In FY 2007, there was a high demand for NETAP training courses, including Procedures for Post Earthquake Safety Evaluation of Buildings (ATC-20); Rapid Visual Screening (RVS) of Buildings for Potential Seismic Hazards (FEMA 154); and Earthquake Hazard Mitigation for Nonstructural Elements. Through these and other courses, FEMA's training target for FY 2007 was surpassed.

In FY 2007, FEMA offered a new training workshop, Earthquake Mitigation for Hospitals, which will serve as a model training program for hospitals on nonstructural mitigation and incremental building rehabilitation. The workshop was pilot-tested in St. Louis in May. Additional training was conducted in Missouri in July with the Department of Mental Health. FEMA also developed a new training course, Seismic Retrofit of Residential Structures, which will be pilot-tested in two venues in Oregon starting in early 2008.

FEMA continued to update and maintain the NEHRP Earthquake Coordinators Web Site. This Web site provides state and federal earthquake coordinators with training on earthquake basics, hazards, risks, building techniques, advocacy and partnerships, and priorities and successful activities (www.training.fema.gov/emiweb/EarthQuake/welcome.htm).

Objective 13: Increase Public Awareness of Earthquake Hazards And Risk

FEMA Web Site for Public Information and Outreach

FEMA's NEHRP Web site (www.fema.gov/hazard/earthquake/ index.shtm), which is linked to the www.nehrp.gov site, includes new sections designed to inform the public, emergency personnel, businesses, and federal, state, and local agencies of ongoing activities in earthquake mitigation by all of the NEHRP agencies and their partners. FEMA continues to post NEHRP technical and non-technical publications in PDF format and text versions on the site, including FEMA 526, Earthquake Safety Checklist, and FEMA 529, Drop, Cover, and Hold, both of which were translated into multiple languages.

Translation of Earthquake Outreach Document

As part of a multilingual earthquake readiness campaign in the San Francisco Bay Area, the popular earthquake preparedness handbook Putting Down Roots in Earthquake Country was translated into Spanish, Chinese, Korean, and Vietnamese. With support from the California Earthquake Authority and PG&E, the USGS and FEMA partnered with the American Red Cross, the California Governor's Office of Emergency Services, New America Media, the Asian Pacific Fund, KTSF 26, and 99 Ranch Market to develop and distribute the new guide, Protecting Tour Familyfrom Earthquakes. Over 700,000 copies of the handbooks were distributed as inserts in 22 local ethnic media publications during the week of the Chinese New Year, when families traditionally prepare for the next year. This release was accompanied by public service announcements and followed by town hall meetings in San Jose and San Francisco in March to raise earthquake awareness by ethnic communities. Since February 2007, these outreach efforts have resulted in requests for 100,000 additional copies.

Multilingual publication on earthquake safety for households. Cover courtesy of the USGS.

Post-Earthquake Information Clearinghouses

Post/earthquake clearinghouses for disaster information have been implemented following major domestic earthquakes for several decades. The clearinghouse concept was adapted in Hurricane Katrina to include multiple phases, including support of immediate planning needs to assist with response and recovery and subsequent support of risk analysis and long/term mitigation activities. FEMA and the USGS are developing protocols and pre/scripted mission assignments to institutionalize clearinghouse events.

CUSEC made significant progress in developing a post/earthquake clearinghouse protocol for the central United States. A special workshop brought together emergency management personnel with representatives from the scientific community to address the varying programmatic and professional needs following an earthquake and how best to coordinate that interaction. This workshop led to the establishment of a clearinghouse committee that will further explore the issues as the planning effort moves forward.

Earthquake Awareness

CUSEC coordinated and supported earthquake awareness programs in Arkansas, Indiana, Kentucky, Missouri, and Tennessee. Support included town hall meetings, presentations at special events, hosting a meeting of the Central United States Seismic Advisory Council, and displays at the St. Louis Children's Museum and the "Earthquakes: Mean Business" forum, which was attended by more than 300 business leaders from the St. Louis area.

NESEC promotes multihazard preparedness and risk reduction to its member states in the Northeast through various media and publications. N ES EC News is published quarterly and provides updates on new state officials and board membership, state activities, success stories, resources, and multihazard risk information. NESEC, which also provides links and information on obtaining GIS software and technical manuals on its Web site, has recently moved its HAZUS-MH/GIS services request system online at www.nesec.org/resources/.

Hayward Fault Alliance

As part of an earthquake preparedness campaign focused on the hazards posed by the Hayward Fault in the east San Francisco Bay Area, the USGS Earthquake Program has partnered with many private/ and public/sector agencies to form the 1868 Hayward Earthquake Alliance. The 1868 Hayward earthquake was the last major earthquake on the Hayward Fault and the Nation's 12th deadliest earthquake. With a magnitude between 6.7 and 7.0, the Hayward earthquake caused 30 deaths and produced extensive damage in San Francisco, San Jose, and communities in between. Recent studies show that the average interval between the past five large earthquakes on the Hayward Fault is 140 ± 50 years. A September 2007 report by the Bureau of Labor Statistics found that over 1.5 million employees with a total annual wage of $100 billion work in the region expected to be most impacted by a Hayward Fault earthquake (based on USGS strong ground motion estimates). More than 125,000 residents are expected to require housing assistance after a Hayward Fault earthquake. Alliance members include the USGS, the CGS, UC-Berkeley, California State University/East Bay, the American Red Cross, FEMA, the California Earthquake Authority, the Governor's Office of Emergency Services, and more than 40 other organizations. The Alliance, which is growing steadily, now has 55 members and an extensive Web site (www.1868alliance.org).There are plans to coordinate the 1868 Alliance activities with the Dare to Prepare campaign in southern California.

USGS geologist with public officials describing the earthquake history of the Hayward Fault as revealed at an excavation site. Photo courtesy of USGS.

Assessment of Earthquake Predictions

The National Earthquake Prediction Evaluation Council (NEPEC) is a federal advisory committee that provides advice and recommendations to the Director of the USGS on earthquake predictions and related scientific research. NEPEC supports the Director's delegated responsibility under the Stafford Act (Public Law 93/288) to issue timely warnings of potential geologic disasters. NEPEC met on two occasions in 2007. In Portland, the Council discussed with American and Canadian scientists the evidence for and nature of episodic nonvolcanic tremor and aseismic creep events in the Pacific Northwest and their possible relation to seismic slip on the Cascadia subduction interface. Since ongoing research suggests that episodes of creep may load the fault closer to failure, the meeting also included discussions with emergency managers on how aseismic slip events and other earthquake/related phenomena in Cascadia should be communicated to the public and on the type of public information most appropriate and effective when episodic tremor and creep events are detected. NEPEC also provided oversight of the review process for the new California earthquake rupture forecast developed by the Working Group on California Earthquake Probabilities.

Objective 14: Develop the Nation's human resource base in earthquake safety fields

NSF Undergraduates Program

In FY 2007, several NSF/supported centers and consortia hosted summer research experiences for undergraduate programs; each typically involved 10 to 20 students working at various institutions. This effort involved SCEC, MCEER, the MAE Center, NEESinc, and the PEER Center. Fourteen student teams competed in the Fourth Annual Seismic Design Competition for Undergraduates, jointly sponsored by the PEER Center, EERI, the MAE Center, and MCEER. The event, held during EERI's Annual Meeting in Los Angeles in February 2007, was an opportunity to demonstrate performance/based, cost/effective seismic design for model structures subjected to earthquake simulation on a small shake table.

Grade-schoolers learn about earthquakes during a visit to the National Earthquake Information Center. Photo courtesy of USGS.

Earthquake Science and High-Performance (Cyber) Computing

The NSF supported SCEC to address the educational challenge of linking earthquake science and cyber-related career pathways through an implementation project, Advancement of Cyberinfrastructure Careers through Earthquake System Science (ACCESS), which is supported under the Cyberinfrastructure Training, Education, Advancement, and Mentoring for Our 21st Century Workforce program. The objective of ACCESS is to provide a diverse group of students with research experiences in earthquake system science that will advance their careers and encourage their creative participation in cyberinfrastructure (CI) development. Its overarching goal is to prepare a diverse, CI/savvy workforce for solving the fundamental problems of system science. ACCESS will encourage women and students from underrepresented and disadvantaged groups to achieve advanced degrees through CI/related research and will guide them toward faculty positions.

USGS Post-Doctoral Program

In FY 2001, the USGS began a program of competitive opportunities for post/doctoral research in the various areas being addressed by its geology programs. Called Mendenhall Fellowships in honor of the Survey's fifth Director, these appointments are for 2 years and provide salary and support for research equipment, data, and fieldwork. Since 2001, 12 scientists have received Mendenhall Fellowships to work at the USGS on research problems in earthquake fields related to the USGS role in NEHRP.

3.4. Develop, Operate, and Maintain NEHRP Facilities

Public Law 108/360 requires that NEHRP "develop, operate, and maintain" certain facilities essential to the NEHRP mission. These facilities are ANSS, NEES, and the Global Seismographic Network (GSN). Reports on the activities and status of these facilities during FY 2007 are given below.

Advanced National Seismic System

ANSS is a national effort led by the USGS to integrate, modernize, and expand seismic monitoring, data analysis, and earthquake notification in the United States. It is based on the cooperative efforts of national and regional earthquake monitoring networks and data analysis centers, including the National Earthquake Information Center (NEIC), which is located in Golden, Colorado. By the end of FY 2007, the ANSS had grown to 784 stations—about 11 percent of the planned 7,100/station network. Dense urban instrumentation has been installed in 5 of 26 at/risk urban centers, and a wide range of products is produced and distributed to stakeholders and the public. ANSS ShakeMaps are seamlessly integrated into FEMA loss estimates, forming the basis for earthquake disaster declarations. ANSS data are channeled to the NOAA Tsunami Warning Centers and to scientific and engineering data centers, enabling engineers to improve building design standards and engineering practices to mitigate the impact of earthquakes.

At the NEIC, performance has dramatically improved since the beginning of 24/7 operations in January 2006, and has further improved as upgrades have been implemented. For significant earthquakes, analyst/reviewed information on location and magnitude is now distributed in less than 20 minutes following earthquake occurrence, an improvement of 50 percent or more from past years. Comprehensive performance standards were set for ANSS this year, which include minimum targets for product delivery timeliness and content, accuracy, and communications. Separate targets are established for the Nation's high/risk urban areas, high/hazard regions, the country as a whole, and areas outside of the United States.

In 2007, ANSS was reviewed by the Investment Review Board of the Department of the Interior (DOI). Among 60 major information technology investments, ANSS ranked highest for business value to the mission of the USGS and DOI and lowest for implementation and operational risk.

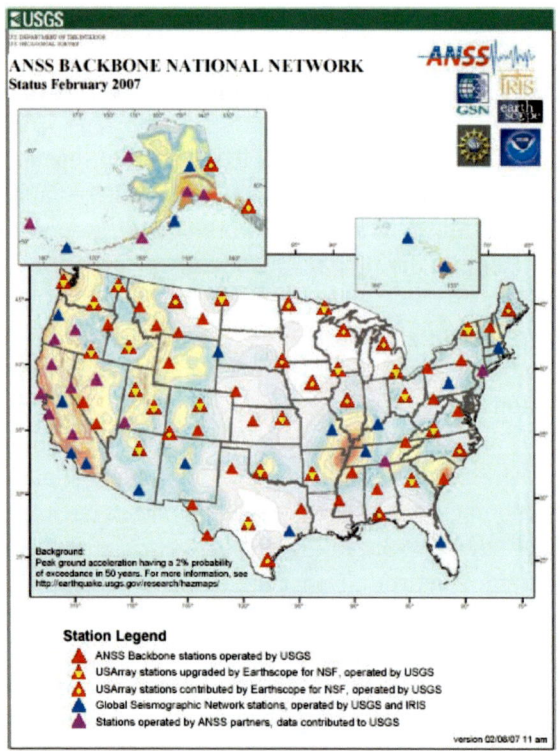

This map shows the stations of the ANSS "Backbone" seismic network, which is largely complete and supports the uniform monitoring and real-time reporting of larger earthquakes in the U.S. The completion of this "Backbone" element of ANSS was possible through partnerships with NSF, NOAA, the U.S. Air Force, and academic institutions. In high hazard areas of the country (background map), the Backbone is supp lemented by dense regional networks, which remain to be modernized under ANSS. Image courtesy of USGS.

This map shows the stations of the ANSS "Backbone" seismic network, which is largely complete and supports the uniform monitoring and real-time reporting of larger earthquakes in the U.S. The completion of this "Backbone" element ofANSS was possible through partnerships with NSF, N0AA, the U.S. Air Force, and academic institutions. In high hazard areas of the country (background map), the Backbone is supplemented by dense regional networks, which remain to be modernized under ANSS. Image courtesy of USGS.

New National Center for Engineering Strong Motion Data

The USGS and the CGS have established a cooperative U.S. National Center for Engineering Strong Motion Data (NCESMD), co/hosted by the CGS and the USGS at www.strongmotioncenter.org. The NCESMD integrates earthquake strong/motion data from the Strong Motion Instrumentation Program of the CGS and the USGS National Strong Motion Project (NSMP) of ANSS. In addition, the NCESMD will assimilate the Virtual Data Center (VDC) developed at UC-Santa Barbara with support from the USGS, SCEC, the NSF, and the Consortium of Strong Motion Observation Systems. The NCESMD Web site provides Internet Quick Reports (IQR), an archive of previous IQRs, and a search engine. The IQRs, which continue to provide the most current strong motion data of engineering significance, are intended primarily for post/ earthquake response and analyses. The first version of an IQR is usually released within a short time (the goal is less than 30 minutes) after an event, and the IQR is updated as additional data are received. A new interactive map feature allows users to view a map of epicenters and strong motion stations using features of the GoogleMaps Web service. By incorporating the VDC, the NCESMD will provide users with a convenient one/stop portal to both U.S. and significant international strong motion data.

Activities and progress of the various ANSS regions in FY 2007 are reported below.

Alaska

A new ShakeMap system is in the final stages of testing at the Alaska Earthquake Information Center (AEIC). A ShakeMap is generated and posted on the AEIC Web site a few minutes after an eligible event is recorded in the database. Overall, there are 7 strong motion, 63 broadband, and 21 broadband+ strong/motion stations contributing into ShakeMap generation at the AEIC. USGS stations located in Anchorage are also automatically incorporated into the ShakeMap production. The AEIC has now incorporated about 120 GSN stations into routine earthquake processing. This considerably improves regional processing by eliminating teleseisms from falsely building into regional events. Eleven new stations were installed, including 8 of 11 planned stations along the

Trans/Alaska Pipeline. Threshold alarms are now being implemented for the Pipeline project.

California

In FY 2007, the ANSS CISN upgraded five analog stations to digital instrumentation with tri-axial accelerometers, converted one USGS National Strong Motion station to continuous recording, and upgraded seven stations to digital telemetry. Seismic data from PG&E's network around Diablo Canyon was integrated into the Northern California Seismic Network (NCSN) in October 2006. In addition, all historic PG&E parametric data were merged with the NCSN phase data and imported into the Northern California Earthquake Data Center (NCEDC) for public use. Real/time waveforms from several of the stations are also transmitted to southern California. In January 2007, PG&E began a project to upgrade its analog seismic stations to digital recording. It has completed the upgrade of six stations, all of which have three acceleration channels and three velocity channels. The data from these upgraded stations are immediately integrated into the NCSN and are used in all CISN products including ShakeMap.

The Southern California Seismic Network (SCSN) and SCEC installed two new transportable stations that transmit real/time data to Pasadena. These stations would be used in post/earthquake response and moved quickly from their existing location to new locations where large aftershocks are expected. In cooperation with the CGS, four strong motion stations along the central San Andreas Fault Zone were upgraded and integrated into SCSN. Southern California also installed three new stations jointly with USArray, which will continue to operate long/term as part of the SCSN/CISN. The CISN also completed the transition statewide to new CISN earthquake processing software.

CISN partners are now working with private companies to assist with seismic monitoring near geothermal plants. A link was established to the earthquake recording system operated by Lawrence Berkeley National Laboratories for the Calpine Corporation at The Geysers geothermal field. This network consists of 23 digital stations with tri-axial velocity sensors digitized at 500 samples per second. The integration of Calpine data into CISN operations enables the NCSN to more accurately locate earthquakes at The Geysers geothermal field. It allows Calpine to put its data into the public domain and relieves them of the need to process and archive the data. It eliminates redundant and discrepant information about earthquakes, which may confuse the public. Finally, it makes an interesting data set available at the NCEDC to seismologists who study triggered earthquake activity, induced seismicity, and geothermal reservoir stimulation. In southern

California, work began to integrate eight new stations in the CalElectric Imperial Valley geothermal field.

Hawaii

The USGS NSMP has completed an upgrade of additional stations on Hawaii Island (Hawaii County) to dial/up communications. There are now 23 NSMP sites in Hawaii County, including a 12/channel building array in Hilo. Of five NSMP stations in Maui County (populated islands of Maui, Molokai, and Lanai), three on Maui are now dial/up. Four stations on Oahu (City and County of Honolulu) remain to be upgraded. The ShakeMap home page has been updated to directly link to Hawaii ShakeMaps, which were previously available only as "Global" events. Hawaii has also begun to investigate and plan presentation of ShakeCast to potential users in the region.

The USGS NEHRP Coordinator convened a committee to assess USGS seismic network capabilities in Hawaii. In part, this was in response to the October 2006 damaging earthquakes that occurred beneath the northwest coast of Hawaii Island. The committee included scientists from the USGS Earthquake and Volcano Hazards Programs and the Pacific Tsunami Warning Center (PTWC). A report submitted to NEHRP includes recommendations for improving implementation, operations, coordination, and management.

The ANSS Hawaii Regional Working Group has agreed to use the CISN organizational model and to form the Hawaii Integrated Seismic Network (HISN). In addition to core partners from NOAA and the USGS, the Hawaii State Civil Defense agency is the third founding member of the HISN. The HISN met for the first time in mid/October 2007.

Pacific Northwest

Infrastructure upgrades are under way in the ANSS Pacific Northwest Seismic Network (PNSN), including upgrades to the production system running in automatic fail/over, several development systems, and special/purpose systems. ShakeMap generation was upgraded with improved geology, and results are now being evaluated; high/resolution geology for Seattle was incorporated in spring 2007. Redundant ShakeMap computers were set up in summer 2007. A new project for the WSDOT, which is now operational, will send a prioritized list of 3,000 vulnerable bridges and their probability of damage to the WSDOT within a few minutes after any earthquake over magnitude 4.

Fifteen stations of the Earthscope/USArray Transportable Array (TA) (see Section 4.1) are being added to PNSN operations through a Murdock Trust grant, as well as three TA stations from Pacific Northwest Laboratory and three stations to be acquired by the State of Oregon. This has allowed the PNSN "authoritative region" to be expanded to encompass the entire States of Washington and Oregon. The USGS also funded the Washington State Office of Emergency Services to provide 15 computers to local jurisdictions to run CISN Display. Six portable broadband instruments are now in use in a training mode, exploring uses for structural monitoring. Three new strong motion recording stations are also in place.

Intermountain West

In Idaho, the Idaho Bureau of Homeland Security and the Idaho Geological Survey have created the Idaho Seismic Advisory Committee to promote seismic hazard mitigation. On September 6–7, 2007, a meeting was held in Boise with about two dozen participants divided into four working groups, including one for seismic monitoring. Key recommendations from the Seismic Monitoring Working Group included undertaking an assessment of what is needed to meet minimum ANSS performance standards in Idaho; building a state funding plan to try to adopt a subset of USArray TA stations; improving coordination with the NEIC and training for emergency managers to understand and use available earthquake information products; establishing the need in Idaho for a state/funded earthquake seismologist; and improving seismic monitoring of rockbursts in the Couer d'Alene hard rock mining district in northern Idaho.

The University of Utah effectively responded to the Crandall Canyon mine/collapse disaster, including media response, field recording, data analysis, and response to investigators. The University is continuing to respond to lawyers and investigators involved in formal investigations being made by Congress, the Mine Safety and Health Administration, and a state/appointed commission. A formal continuity of operations plan has been completed, and the University is also working with one/time funding from the state to develop redundant recording outside of the network operations center and back/up telemetry routing to the NEIC in case of a large Wasatch Front earthquake. Using state funds, a 36/channel array is being installed in the Utah State Capitol for seismic structural/response monitoring.

Northeast United States

ANSS-Northeast is implementing the ShakeMap software at Columbia University's Lamont/ Doherty Earth Observatory. Boston College's Weston Observatory has developed geologic maps for the region to obtain soil amplification. Studies of the 2006 Maine earthquake sequence have helped delineate the structure upon which that earthquake sequence took place. The

Roundtable of ANSS-Northeast Stakeholders was held September 17–18, 2007, at the Boston College Conference Center. There were 38 participants, including representatives from most of the 12 states in the region. Attendees included state geologists, emergency managers, earthquake engineers, seismologists, federal representatives, and private industry.

Central United States
A borehole seismic array was installed on the east bank of the Mississippi River near the Interstate 40 crossing in downtown Memphis. The array consists of three broadband and three strong motion channels at the surface, 31/meter depth, and 62/meter depth. The array is currently in local triggered mode only. An agreement was entered into with the Tennessee Department of Transportation (TDOT) to send data back to the University of Memphis Center for Earthquake Research and Information via the TDOT "Smart Highway" wireless/fiber network (scheduled for 2008).

George E. Brown, Jr. Network for Earthquake Engineering Simulation
On September 30, 2004, NEES completed its 5/year, $82 million major research equipment and facilities construction. The 15 NEES experimental facilities, located at academic institutions across the United States, include shake tables, geotechnical centrifuges, a tsunami wave basin, large strong floor and reaction wall facilities with unique testing equipment, and mobile and permanently installed field equipment. Through the NEES information technology CI, these 15 experimental facilities are linked via the Internet2 grid, forming the world's first prototype of a distributed "virtual instrument," and can be connected with similar facilities worldwide to harness the best talent globally for earthquake engineering research.

NEES operations and use of its facilities for research and education began on October 1, 2004, under management by NEES Consortium, Inc. (NEESinc), located in Davis, California. NEESinc is a nonprofit organization that works in partnership with the 15 universities to operate the NEES experimental facilities and CI. NEESinc manages NEES as a national, shared/use resource for research and education for the earthquake engineering community and schedules access to the experimental facilities. NEESinc also provides the system/wide information technology infrastructure of NEES, including repositories for NEES data and simulation tools; manages an education, outreach, and training program; and fosters linkages and partnerships with federal, state, and local government entities, national laboratories, the private sector, and international collaborators. During the summers of 2006 and 2007, NEESinc ran research experiences for undergraduate programs that gave students the opportunity to work at the NEES experimental facilities (www.nees.org).

The NEES Nonstructural Component Simulator. Image courtesy of the University at Buffalo's Structural Engineering and Earthquake Simulation Laboratory (SEESL).

NEES provides unique opportunities to pursue the high/priority research outlined in the 2003 National Research Council report, Preventing Earthquake Disasters— The Grand Challenge in Earthquake Engineering; to demonstrate the validity of seismic design and rehabilitation concepts; to speed the transfer of research into seismic design guidelines and specifications; and to develop well/informed disaster preparedness and recovery strategies. The NEES infrastructure (experimental facilities and CI) facilitates a variety of innovative experimental approaches that are leading to a better understanding of how the built environment, e.g., buildings, bridges, utility systems, coastal regions, and geomaterials, performs during seismic events. Through four annual program solicitations and the Small Grants for Exploratory Research program, the NSF has funded over 40 research projects to apply the NEES facilities to study soil foundation and structure interaction; seismic performance of foundations, lifelines, and reinforced concrete, masonry, wood, and composite structures; behavior of steel frames with innovative bracing schemes; seismic design of nonstructural systems; seismic risk mitigation of ports; and the seismic performance of bridge systems with conventional and innovative materials. Many of these projects include practitioner and industry partners to help design experimental and analytical investigations and to speed technology transfer. NEES also provides national resources for developing, coordinating, and sharing new educational programs and materials to train the next generation of the earthquake engineering workforce.

Model bridge tested at the University of Nevada, Reno (UNR) NEES facility, February 2007. Image courtesy of UNR.

Global Seismographic Network

The GSN is a worldwide network of seismic recording stations with standardized instrument design, data format, and communication protocols. The GSN is a joint program implemented by the USGS through its Albuquerque Seismological Laboratory and the NSF through Incorporated Research Institutions for Seismology and the Institute of Geophysics and Planetary Physics of the University of California.

New GSN Stations

In FY 2007, the number of GSN stations increased from 141 to 147. In the Republic of Kiribati, a new station was installed on Tarawa Island; the GSN station on Christmas Island was restored to an operational status; and with the assistance of the U.S. Coast Guard, equipment was pre/deployed to the atoll of Kanton in anticipation of installation in the fall 2007. A new GSN station was also installed at Ambohimpanompo, Madagascar. In other areas, equipment to reinstall the station on Wake Island (destroyed by a hurricane last year) has been shipped. Site work has started on new stations in Baja California, the Canary Islands, and the United Arab Emirates. These stations will be operational early in 2008. Field teams also performed maintenance on stations in Chile, Ecuador, Mali, Ethiopia, the Cook Islands, the Solomon Islands, Texas, and elsewhere.

Caribbean Seismic Network Expansion (a GSN Affiliate Network)

Of nine stations planned for the Caribbean Seismic Network, seven were fully installed and operating by the end of 2007. These stations are located at sites in Antigua/Barbuda, the U.S. naval base at Guantanamo Bay; the Dominican Republic; Grenada, Barbados; Honduras; and Panama. All seven are transmitting real/time data that have been integrated into NOAA tsunami warning centers and USGS/NEIC operations. Equipment was shipped for the remaining two stations of the Caribbean Network (Jamaica and Turks and Caicos), which are expected to transmit data by the end of 2007.

Real-Time Data Communications Upgrades

NEHRP continued to expand and improve telemetry to GSN stations. Improved communications capabilities have been achieved at 36 of 39 planned sites. At 21 sites, bandwidth has been enhanced by upgrade of the VSAT system supported by the Comprehensive Test Ban Treaty Organization; an additional 8 sites have new communications links and 4 sites of the Pacific satellite system were installed. The USGS and NOAA are collaborating for the upgrade of telemetry from stations to PTWC and onward to the NEIC. NOAA and the USGS added five stations to the satellite communications system in 2007 and will add two additional stations by 2008.

4. RELATED ACTIVITIES SUPPORTING NEHRP GOALS

Public Law 108/360, the Earthquake Hazards Reduction Program Reauthorization Act of 2004, requires that the annual report to Congress include a description of activities being carried out by the National Earthquake Hazards Reduction Program (NEHRP) agencies that contribute to NEHRP goals but are not officially included in the Program. Highlights of these programs and activities are described below.

4.1. Earthscope

EarthScope is a multipurpose array of instruments and observatories that will greatly help to advance understanding of the structure, evolution, and dynamics of the North American continent. The program, which is supported by the National Science Foundation (NSF) in partnership with the U.S. Geological Survey (USGS), the National Aeronautical and Space Administration (NASA), and academic institutions, provides an integrative framework for research on fault properties and the earthquake process, strain transfer, magmatic and hydrous fluids in the crust and mantle, plate boundary processes, large/scale continental deformation, continental structure and evolution, and composition and structure of the deep/Earth. In addition, EarthScope offers a centralized forum for earth science education at all levels and an excellent opportunity to develop cyberinfrastructure (CI) to integrate, distribute, and analyze diverse data sets.

The primary elements of the program are the San Andreas Fault Observatory at Depth (SAFOD), the Plate Boundary Observatory (PBO), and the USArray. These components are constructed, operated, and maintained as a collaborative effort with UNAVCO, the Incorporated Research Institutions for Seismology, the USGS, NASA, Stanford University, and several other national and international institutions. In 2007, the deployment of instrumentation included approximately 870 Global Positioning System (GPS) instruments, 485 seismometers, 35 strain meters, and 108 magnetotelluric instruments.

SAFOD drill rig, July 2004. Photo courtesy of EarthScope-SAFOD.

San Andreas Fault Observatory at Depth

Phase 3 of the SAFOD drilling program began in the summer 2007. This stage involves multiple drillings from the main hole to obtain continuous cores within the San Andreas Fault Zone. Researchers have retrieved 10.67 meters (135 feet) of rock core across the actively slipping fault zone at a depth of 2.2 kilometers. Researchers hope to use the core to address several outstanding questions about the fault's properties and materials, as well as earthquake generation. These questions include how the mechanics of the creeping section of the San Andreas Fault differ from those sections that cause episodic, large earthquakes, and how different minerals facilitate motion along the fault interfaces. The science team will install a host of seismic instruments in the 4.02 kilometers (2.5-mile)-long borehole that runs from the Pacific Plate on the west side of the fault into the North American Plate on the east. By placing sensors next to a zone of small repeating earthquakes, scientists will be able to observe the earthquake generation process in unprecedented ways.

SAFOD has drilled into this fault zone to study the physics of earthquake nucleation and rupture and determine the composition, physical properties, and mechanical behavior of an active, plate- bounding fault at seismogenic depths. SAFOD, which is located near Parkfield, California, penetrates a section of the fault that is moving through a combination of repeating microearthquakes and fault creep. Earlier drilling in the summer 2004 and 2005 went down vertically to a depth of 1.5 kilometers and then deviated to penetrate the active San Andreas Fault Zone at a vertical depth of about 2.7 kilometers. In the summer 2007, cores were acquired from holes branching off the main SAFOD borehole to directly sample fault and country rocks at depth. The observations from these holes define the San Andreas Fault Zone as relatively broad (around 250 meters), containing several discrete, actively creeping shear zones that were targeted for coring. The occurrence of the mineral serpentinite in the actively deforming zones is particularly significant because it is widely regarded to be important in controlling frictional strength and the stability of sliding. The core samples will be extensively tested in the laboratory to study the composition, deformation mechanisms, physical properties, and rheological behavior of fault rocks from the active traces of the San Andreas Fault at realistic in/situ conditions. The results provide realistic values for use in computer models of earthquake processes.

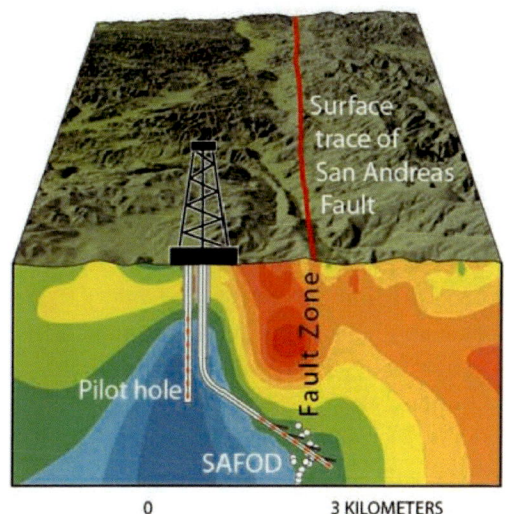

Schematic cross section of the San Andreas Fault Zone at Parkfield, showing the SAFOD drill hole and a pilot hole drilled in 2002. Red dots in holes show sites of monitoring instruments; white dots represent area of persistent minor seismicity. Illustration courtesy of Stanford University.

Plate Boundary Observatory

The PBO is the geodetic component of EarthScope that will facilitate the study of deformation across the active plate boundary zone between the Pacific and North American Plates along the western United States. From January 14, 2007, to February 1, 2007, GPS, strain/meter, and seismic instrumentation captured an episodic tremor and slip (ETS) event which began under the southern Puget Sound area to the southwest of Seattle and propagated north northwest into Vancouver Island, Canada, at a rate of 10 kilometers per day. Researchers estimate that enough energy was released in this 3/week episode to equal a magnitude 6.6 earthquake. These ETS events can affect the magnitude and timing of future large earthquakes within the subduction zone and thus are important in advancing understanding of seismic hazards in the region.

Data collection for the GeoEarthScope Northern California Light Detection and Ranging (LiDAR) project concluded in April 2007. This section of the San Andreas Fault lies within the greater San Francisco Bay Area, making it among the most important in the United States in terms of seismic hazard. LiDAR is particularly useful for imaging the region because of the forest cover that hampers other imaging techniques. Over the course of several weeks, approximately 1,400 square kilometers of EarthScope targets were imaged as well as supplementary targets for the USGS. Between this project and the NSF/funded B4 project, the entire San Andreas Fault system, along with many other important adjacent faults and structures, have now been imaged with high/ resolution airborne LiDAR.

USArray

Under USArray, permanent seismic stations have been installed in partnership with the USGS to help complete an equally spaced network of seismic stations that can detect and record earthquakes of magnitude 3.0 or greater across the United States. Intermediate/term (up to 2 years) and short/ term Transportable Array (TA) stations are providing even more detailed earthquake, earth structure, and ground shaking information.

Installation of the TA network is complete in California, Oregon, Washington, and Nevada, with stations located approximately 70 kilometers apart in a gridlike fashion. Coverage is nearly complete in Arizona, with installation of only a few more stations required. While construction crews have been busy in Idaho, Utah, and western Montana, site reconnaissance activities are in full swing in New Mexico, Colorado, Wyoming, central and eastern Montana, and in the Big Bend area of Texas. For the third year in a row, student teams from local universities were trained to perform analytical and field activities to select suitable locations for future seismic stations. In the summer 2007, more than 150 sites were successfully identified by eight ambitious teams from Colorado College, the University of Wyoming, Montana Tech, and the University of Texas/El Paso.

4.2. Subcommittee on Disaster Reductio

Many federal agencies play important roles in reducing the effects and impacts of natural hazards. The White House National Science and Technology Council's Committee on Environment and Natural Resources, Subcommittee on Disaster Reduction (SDR) provides coordination for the full spectrum of science and technology contributions to disaster reduction. The SDR is charged with establishing national goals for federal science and technology investments in disaster reduction. In support of this mission, the SDR provides a senior/level interagency forum to leverage expertise, inform policy makers, promote technology applications, coordinate activities, and promote excellence in research and development. NEHRP agencies are all actively involved in SDR activities and serve as the focal point for earthquake/related topics, ensuring coordination between NEHRP and the broader federal enterprise.

4.3. International Activities

U.S.-Japan Program on Natural Resources
In 1964, the United States and Japan established the bilateral U.S./Japan Program on Natural Resources (UJNR) to promote bilateral cooperation in research and data exchange. Today, the UJNR involves 18 U.S. agencies and 10 Japanese agencies. The NEHRP member agencies play important roles in the UJNR panels on wind and seismic effects and on earthquake research. The U.S. sides of the panels are chaired by the National Institute of Standards and Technology and the USGS, respectively.

U.S.-Japan Panel on Wind and Seismic Effects
The work of this Panel has resulted in improved building and bridge standards and codes and design and construction practices in hydraulic structures in both countries. The Panel's work involved exchanging guest researchers who performed short/ and long/term joint cooperative research assignments, visiting major public works construction projects that employ innovative civil engineering techniques and research laboratories with unique test and measurement capabilities, and performing joint post/disaster surveys. The Panel's

accomplishments serve as technical bases for improving seismic design and construction practices by advancing retrofit techniques for bridge structures, addressing shake table modeling and simulation of nonlinear problems, assessing earthquake risk of dams, testing seismic performance guidelines for bridge piers, and jointly conducting full/scale column tests at the National Research Institute for Earth Science and Disaster Prevention's (NIED) 3-D Full-Scale Earthquake Testing Facility (E-Defense) in Miki, Japan. Other activities have included producing full-scale test data that advance seismic design standards for buildings, advancing technology for repairing and strengthening reinforced concrete, steel, and masonry structures, improving in-situ measurement methods for soil liquefaction and stability under seismic loads, and creating a database comparing Japanese and U.S. standard penetration tests to improve prediction of soil liquefaction.

U.S.-Japan Panel on Earthquake Research

The UJNR Panel on Earthquake Research promotes advanced research toward a more fundamental understanding of earthquake processes and hazard estimation. The Panel promotes basic and applied research to improve our understanding of the causes and effects of earthquakes and to facilitate the transmission of research results to those who implement hazard reduction measures. The sixth Panel meeting was held in Tokushima, Japan, in November 2006. The meeting included very productive exchanges of information on approaches to systematic observation of earthquake processes. Sixty-eight technical papers were presented during the meeting on a wide range of subjects, including interplate earthquakes in subduction zones, slow slip and nonvolcanic tremor, crustal deformation, recent earthquake activity, and hazard mapping. Through discussions, scientists from the two nations reaffirmed the benefits of working together to achieve a common goal of reducing earthquake hazard, continued cooperation on issues involving densification of observation networks, and the open exchange of data among scientific communities. Participants also reaffirmed the importance of making information public in a timely manner. The Panel visited sites along the east coast of Shikoku that were inundated by the tsunami caused by the 1946 Nankai earthquake, hearing from survivors of the disaster and seeing new tsunami shelters and barriers. They also visited the Median Tectonic Line, a major onshore strike-slip fault on Shikoku.

U.S.-China Cooperation in Earthquake Studies

In fiscal year 2007, there were numerous bilateral interactions between U.S. and Chinese investigators covering earthquake hazard reduction topics. Common interests included enhanced seismic and GPS monitoring, improved rapid notification, research, and public education. A highlight of the year was a visit in May 2007 to the China Earthquake Administration in Beijing by a 17-person U.S. delegation led by representatives of the NSF and the USGS. The purpose of this visit was to review progress achieved in past bilateral activities and to plan new joint work. The 3-day meeting was attended by leading Chinese earthquake scientists and administrators, and included a visit to newly established research and seismic monitoring facilities in Beijing. An outcome of the visit was the decision to hold an open, 4-day bilateral workshop in the United States in May 2008 on recent progress in earthquake studies. This workshop will be cosponsored by the NSF and the USGS. There are numerous ongoing bilateral activities as well. Prominent examples of these are the exchange of broadband seismic data (unfortunately not in real time) and GPS processed data. Such data are critical to understanding the current tectonics of China. There is increased interest in Chinese strong motion data by U.S. investigators, and these data are being made available. All U.S. seismic and GPS data are available to the Chinese (and the world) and much of the U.S. seismic and GPS data are available in real time. The future of U.S./China bilateral activities in earthquake studies is certain to be characterized by a rapid expansion and deepening of cooperation. A major factor driving this expanded cooperation is the large fiscal investment being made by the Chinese in their monitoring systems, research facilities, and staff.

Seismic Hazard Assessment for Afghanistan

The Islamic Republic of Afghanistan is located in a geologically active part of the world, and each year the country is struck by moderate to strong earthquakes. Every few years, a powerful earthquake causes significant damage and/or fatalities. As Afghanistan rebuilds following decades of war and strife, new construction and development should be designed to accommodate the strong shaking and related hazards posed by strong earthquakes. To assist in the reconstruction efforts, the USGS has developed a preliminary earthquake hazard map of Afghanistan, which incorporates data from thousands of historical earthquakes, information about active faults, and models of how earthquake energy travels through the Earth's crust to define expected levels of ground shaking throughout the country (http://pubs.usgs.gov/of/2007/ 1137/).

Even though the seismic hazard map is preliminary, it provides government officials, engineers, and private companies interested in investing in Afghanistan's growth with crucial information about the location and nature of seismic hazards; this information allows them to make informed decisions about the designs and locations of critical structures such as power plants, dams, pipelines, hospitals, and similar facilities.

The map was formally released on May 30, 2007, at a ceremony and news conference at the Afghanistan Embassy in Washington, D.C., where Ambassador H.E. Said Tayeb Jawad presided over the event.

In 2007, the USGS, in cooperation with Kabul University and the Afghan Geological Survey, reestablished the Kabul seismic station (KBL) after a 20/year hiatus. This broadband station is one of a few modern seismograph stations in the region, and data from it provide important new details about the location, size, and depth of earthquakes throughout Afghanistan and southern Asia. Real/time data from the KBL station is an integral part of the USGS's Global Seismographic Network (GSN).

EES/E-Defense

The George E. Brown, Jr. Network for Earthquake Engineering Simulation (NEES) is leveraging and complementing its capabilities through connections and collaborations with large testing facilities at foreign earthquake/related centers, laboratories, and institutions. The NSF and the NEES Consortium, Inc. (NEESinc) have recently developed partnerships to utilize the NEES infrastructure with E-Defense of NIED, which became operational in 2005. To facilitate NEES/E-Defense collaboration, NEESinc and NIED signed a memorandum of understanding in August 2005. In September 2005, the NSF and the Japanese Ministry of Education, Culture, Sports, Science, and Technology signed a memorandum concerning cooperation in the area of disaster prevention research. Through such partnerships and joint meetings and workshops, NEES shares its expertise in testing and CI, provides specialized training opportunities, and coordinates access to unique testing facilities and the central data repository. Five NSF/supported research projects addressing the seismic performance of bridge columns, mid/rise wooden buildings, steel frames, and base/ isolated structures will utilize both NEES facilities and E-Defense in the conduct of their projects during the 2008–2010 timeframe.

U.S. Contribution to GEOSS

The GSN has been contributed by the United States to the Global Earth Observation System of Systems (GEOSS). GSN and International Federation of Digital Seismograph Networks seismic stations provide broad, extensive coverage over all continents and on oceanic islands. Data availability of the GSN is better than 85 percent, and more than 90 percent of stations are openly available in real time. Data availability in many national networks approaches 99 percent and data availability through the international data archiving system is rapidly improving.

4.4. NEHRP Contributions to Tsunami Safety

Tsunami Hazard Assessment for the United States
In 2005, the U.S. National Science and Technology Council released a joint report by the SDR and the U.S. Group on Earth Observations, Tsunami Risk Reduction or the United States: A Framework or Action. The first specific action called for in the Framework is to "Develop standardized and coordinated tsunami hazard and risk assessments for all coastal regions of the United States and its territories." The National Geophysical Data Center (NGDC) of the National Oceanic and Atmospheric Administration (NOAA) and the USGS partnered to conduct the first tsunami hazard assessment for the United States and its territories. The U.S. States and Territories National Tsunami Assessment: Historical Record and Sources or Waves (in press, 2007) is the first step toward a national tsunami risk assessment. The goal of the report is to provide a qualitative assessment of the U.S. tsunami hazard at the national level.

The core of the assessment involved dividing the NGDC historical tsunami database based on the measured sea level run/up heights and the number of run/ups at each height. The second assessment method used USGS estimates of recurrence of possible tsunami/generating earthquakes near U.S. coastlines to extend the NOAA/NGDC tsunami databases back in time. Combining the two techniques shows that the U.S. tsunami hazard is highest for all Pacific states and possessions, and for those possessions in the Caribbean. The history of large run/ups causing extensive destruction and fatalities in Alaska (e.g. Unimak 1946, Valdez 1964) and Hawaii (e.g. Hilo 1946, 1960) indicate that the tsunami hazards there are very high. In contrast, hazards are low to very low along the Atlantic seaboard and very low for the gulf coast states.

Global Training in Seismology and Tsunami Warning
The Indian Ocean Tsunami Warning System development program, of which the USGS is a part, is beginning its closeout phase. Activities over the past 2 years have enhanced the ability of government agencies in the region to assess and respond to their risk from earthquake and tsunami hazards. These activities included (1) technical discussions with partners on warning system design, institutional capacity building, and system integration; (2) characterizing regional seismic hazards and incorporating updated hazard assessments into building

codes; (3) upgrading the GPS and seismic network of Indonesia, including the introduction of real/time telemetry; and (4) maximizing and ensuring the future sustainability of the seismic and tsunami programs with regional partners.

To support capacity that has expanded since the Andaman/Sumatra earthquake and tsunami of 2004, and to build a base of expertise in seismology and tsunami warnings in countries where these topics were relatively new, the USGS, in partnership with the U.S. Agency for International Development, the United Nations Educational, Scientific and Cultural Organization's Intergovernmental Oceanographic Commission, and NOAA, continued its training program in earthquake monitoring and tsunami warnings for national warning centers in southeast Asian nations surrounding the Indian Ocean region. Four courses were completed in the Indian Ocean region last year, training operational staff from several countries, including Thailand, Malaysia, Indonesia, Sri Lanka, The Maldives, and Vietnam. Additional training in observatory practice and data management was conducted at the USGS National Earthquake Information Center and Menlo Park facilities.

The groundwork and knowledge base to understand and implement a warning system came to fruition with the performance of Indian Ocean nations in response to the magnitude 8.4 and 7.9 southern Sumatra earthquakes of September 12, 2007. Progress in building a warning system that quickly and effectively alerted populations was evident and acknowledged in many newspaper reports.

A region/wide Caribbean training course in seismology and tsunami warning held in July 2007 was hosted by the Seismic Research Unit of the University of the West Indies in Trinidad. The course provided training to approximately 20 to 25 scientists.

NEES Tsunami Wave Basin at Oregon State University

One of the 15 experimental facilities within NEES is the world/class tsunami wave basin at Oregon State University. This shared/use facility is currently being used by researchers from around the United States, through several NSF/supported research projects, to improve understanding of tsunami generation and impact on the built environment and to develop performance/based tsunami engineering.

5. STATE ACTIVITIES TO PROMOTE IMPLEMENTATION OF RESEARCH RESULTS

Through the Mitigation and Preparedness Directorates, the Federal Emergency Management Agency (FEMA) supports state and local efforts to reduce their risks to all hazards, including earthquakes. In addition to development of technical assistance and guidance documents, FEMA administers several competitive and formulaic grant programs, including the all-hazards Pre-Disaster Mitigation (PDM) Grant Program for states and communities; the Hazard Mitigation Grant Program (HMGP), an all-hazards post-disaster grant program; and the Emergency Management Performance Grants (EMPG) Program, which is a FEMA program that provides grants to states to improve emergency management performance and is administered by FEMA's Preparedness Directorate. With these grants, state and local agencies can fund planning activities and projects to protect their citizens from the earthquake hazard. Highlights of successful state, territorial, and local government efforts in support of NEHRP in fiscal year (FY) 2007 are described below.

Alaska

Education and outreach tools used by Alaska include the "Quake Cottage" earthquake simulator and "Earthquake-Resistant Model Home," which support nonstructural seismic hazard mitigation and preparedness outreach demonstrations at statewide venues. Alaska also printed 2,000 copies of Molly and the Earthquake and Heidi and the Tsunami, books which tell fictional stories of a family's natural hazards experience and give safety tips on what to do before, during, and after an earthquake or tsunami.

The National Oceanic and Atmospheric Administration (NOAA), University of Alaska Fairbanks/Geophysical Institute (UAF/GI), and state Division of Homeland Security and Emergency Management sponsored tsunami inundation mapping projects for response and planning efforts in the communities of Homer and Seldovia. UAF/GI is researching future mapping initiatives for potential tsunami events in Seward, Sitka, and Valdez.

Alaska continues to support the statewide post-disaster damage assessment training program using a modified Applied Technology Council (ATC)-20 training course managed by the Municipality of Anchorage's Building Safety Officer. Alaska has approximately 650 trained and certified damage assessors.

Alaska received two FEMA Pre-Disaster Mitigation-Competitive (PDM-C) grants for 2006 (funds were not obligated until 2007). Schools in the Kodiak

Island Borough will receive seismic retrofitting and 10 Municipality of Anchorage schools will receive automatic gas shut-off valve retrofits.

California

The Earthquake Country Alliance (ECA) is a public-private partnership led by the USGS, the City of Los Angeles Emergency Preparedness Department, and the Southern California Earthquake Center. A major ECA project is the 2007 Earthquake Readiness Campaign, Dare to Prepare, to raise earthquake awareness and encourage earthquake readiness in southern California. Components of the Dare to Prepare campaign include the www.daretoprepare.org Web site; distribution of material such as Putting Down Roots in Earthquake Countr~; hosting local public events; and planning the 2008 Great Southern California Shakeout, a regional public earthquake exercise. A second major ECA activity is the Southern San Andreas Fault Evaluation (SoSAFE) project. The objective of SoSAFE is to obtain new data to clarify and refine relative hazard assessments for each potential source of a future "Big One." The first SoSAFE workshop, held on January 9, 2007, marked the sesquicentennial of the great 1857 Fort Tejon earthquake. A reenactment of the 1857 event will be part of the 2008 Great Southern California Shakeout disaster exercise.

The Northern California 1868 Hayward Earthquake Alliance is planning activities to commemorate the 140th anniversary of the 1868 Hayward earthquake. The Alliance's goals include raising public awareness of the imminent hazard posed by the Hayward Fault and reducing its risk to Bay Area residents, communities, and businesses. More information is available at www.1868alliance.org.

Hawaii

On October 15, 2006, a magnitude 6.7 earthquake struck the northwest coast of the Island of Hawaii. The Kiholo Bay earthquake caused severe damage, particularly to the port facilities at Kawaihae Harbor. Given the vulnerability of the port to seismic events and its importance to the economic viability of the island, FEMA Region IX initiated an evaluation of the damage to the port and an assessment of potential mitigation measures. The assessment will provide recommendations to reduce the risk of damage to Kawaihae Harbor and to similar ports in the Pacific. The Kawaihae Harbor evaluation report will be finalized in early 2008.

The magnitude 6.7 earthquake also provided a unique, perishable opportunity to compare Hazards U.S.-Multihazard (HAZUS-MH) analysis results with damage observed in the disaster. FEMA Region IX initiated a project to incorporate the HAZUS-99 inventory, hazard, and modeling improvements completed by Hawaii State Civil Defense into the latest HAZUS-MH (MR-3) platform and use the observed impacts associated with the Kiholo Bay earthquake to validate the modeled results.

Idaho

For more than 20 years, the Idaho Geological Survey (IGS) has encouraged excellence in earth science education by providing secondary and elementary teachers with unique, hands/on, "learning through inquiry" instruction on geology at field locations throughout Idaho. The week/long IGS workshops, which received a 1996 Award in Excellence from the Western States Seismic Policy Council, provide about 30 Idaho educators each year with short courses in geological concepts and methods and expose the educators to new technologies and theories. IGS staff members teach most of the course. The Idaho Bureau of Homeland Security and FEMA, which provide funding for the workshops, offer training in hazard mitigation and pre/disaster planning. Other workshop cosponsors are the National Energy Foundation and the Idaho Mining Association.

The HMGP has funded nonstructural seismic mitigation projects at the Idaho State Controller's Office and the Idaho State Emergency Operations Center. Idaho continues to deliver rapid visual screening (RVS) training to assist local jurisdictions in identifying potential seismic mitigation projects. Idaho also delivered several ATC-20 training courses to build its damage assessment capability.

Illinois

Illinois received an HMGP grant for Waterloo Community Schools District #5 for seismic/ resistant construction of a new high school in March 2007. Through the work of the Mid/America Earthquake Center, the state also completed comprehensive seismic loss modeling for southern portions of Illinois. Illinois conducted point/of/distribution exercises in southern Illinois counties for the New Madrid Seismic Zone (NMSZ) and set a schedule for NMSZ workshops in FY 2008.

Indiana

Indiana distributed the "Preparing for Indiana's Earthquake Risk" video and continued work on seismic gas shut/off valve installation at St. Vincent's Hospitals in Warren County (Williamsport) and Clay County (Brazil). Indiana held three HAZUS-MH basic training classes on earthquake modeling and performed HAZUS-MH MR2 Level 2 Earthquake modeling runs for 33 counties, using data for the New Madrid, Wabash Valley, and Anna Seismic Zones. Indiana began work on its NMSZ Response Plan and will begin NMSZ workshops in the fall of 2007 that will continue in FY 2008.

Kentucky

A full/scale Kentucky National Guard exercise, held in March 2007, took place in Areas 1–3 of western Kentucky and involved a New Madrid earthquake scenario. Guardsmen, emergency management personnel, and the Civil Air Patrol took part in the exercise, along with local agencies from selected counties. Areas 1, 4, and 12 in Kentucky also participated in the Spills of National Significance (SONS) '07 exercise that focused on a New Madrid earthquake. Kentucky hosted a 3-day exercise that covered three areas, including the state Emergency Operations Center. The Shell Oil Company set up a federal portion of the SONS exercise in Paducah, which included local involvement. Pendleton County also hosted an earthquake exercise in 2007.

For FEMA's NMSZ catastrophic planning initiative, Kentucky has hosted three local planning workshops; a state workshop is scheduled for March. This is a much-needed, exciting effort for Kentucky. Significant work is being put into this plan, including the formation of a formal "area command" concept.

Plans for Earthquake Preparedness Week, which will take place in February, include an Earthquake 101 class in western Kentucky in conjunction with the Central United States Earthquake Consortium (CUSEC), school drills, and a teacher curriculum. The CUSEC Public Information Committee is creating a press kit for the event.

Missouri

Two earthquake presentations at the Missouri State Emergency Management Agency/Missouri Emergency Preparedness Association Conference reached more than 100 participants. An "Earthquake Mitigation for Hospitals" poster was presented at the Natural Hazards Workshop in Boulder, Colorado, and at the Earthquake Conference at Rice University in Houston, Texas. Missouri Earthquake Awareness Week 2008 included "Earthquakes: Mean Business," a seminar for business and industry, attended by about 250 participants; the St. Louis Science Center Earthquake Awareness Day, attended by about 350 participants; and the Earthquake Town Hall Meeting at New Madrid, attended by more than 50 participants.

Nevada

The Nevada Earthquake Safety Council (NESC) met with the Utah Seismic Safety Commission (USSC) in St. George, Utah, to forge a working relationship. The joint meeting provided many opportunities for collaborative actions to reduce shared seismic risk. The NESC also participated in the 3-year review and update of the Nevada State Mitigation Plan. The NESC spearheaded the integration of updated seismic hazard information and mitigation action items into the plan, which is under review by FEMA Region IX.

The Nevada Bureau of Mines and Geology Open-File Report 6-1, "Loss Estimation Modeling of Earthquake Scenarios for Each County in Nevada Using HAZUS-MH," won an award at a recent geographic information system conference and is part of the NESC use of HAZUS to heighten awareness of earthquake hazards throughout Nevada.

Oregon

Oregon Emergency Management (OEM) and the Oregon Department of Geology and Mineral Industries (DOGAMI) have collaborated on several seismic safety activities. The Oregon Seismic Safety Policy Advisory Commission supported passage of key legislation to form a Seismic Rehabilitation Grants Committee that will oversee state bond measure funds for strengthening public schools and police, fire, and acute/care emergency facilities. Seismic rehabilitation demonstration projects funded by the PDM grant program are in progress at two universities. DOGAMI issued a statewide risk assessment report and database for 3,300 public schools and emergency facilities. Oregon also mapped earthquake hazards and conducted HAZUS risk assessments for 11 counties for its PDM-funded hazard mitigation plans; mapped geology and faults in areas of Portland; completed an enhanced RVS method and shared it with FEMA-funded ATC-67 project members; and coordinated with the Oregon Public Utility Commission to start seismic audits of energy providers. OEM supported the annual "Earthquake and Tsunami Awareness Month" in April and conducted the Pacific Peril '06 U.S. Department of Transportation exercise in Oregon for a Cascadia earthquake and tsunami scenario.

Puerto Rico

Seismic Activity is an educational campaign for the general public that presents the first earthquake simulator in Puerto Rico and the Caribbean; people can see, feel, and listen to earthquake effects in three different scenarios: at school, at home, and at work. From April 12 to May 20, 2007, events were held at six large shopping malls. Coverage was provided through several media outlets in the island, including live transmission from the shopping malls and public announcements that resulted in the participation of 100,000 citizens. Starting in 2005, this educational campaign has reached at least 80,000 people each year.

The Puerto Rico Emergency Management Agency (PREMA) distributed NOAA radios and a multihazard interactive CD in 984 public schools to generate public awareness among children and schoolteachers. PREMA also conducted 53 earthquake and tsunami workshops, reaching a total of 12,910 citizens. In addition, PREMA offered two earthquake train/the/trainer seminars in which 120 instructors were certified.

South Carolina

In coordination with FEMA, South Carolina offered two courses on earthquake safety (ATC-20 training and FEMA 154-RVS training) that provided 40 attendees with basic principles of nonstructural and structural mitigation measures for earthquakes.

The South Carolina Emergency Management Division (SCEMD) awarded a new South Carolina Earthquake Education and Preparedness Program to the College of Charleston. The College of Charleston's Department of Geology and Environmental Geosciences is recognized as a center of excellence on earthquakes, earthquake education, and earthquake preparedness.

South Carolina's Earthquake Awareness Week was held in November. In support of the week, a governor's proclamation was released, news releases were issued, and earthquake literature was mailed to county emergency managers and schools. A "Drop, Cover, and Hold" drill, announced over the South Carolina NOAA weather radio, was also held with schools. The State Earthquake Plan continues to be updated with HAZUS-based damage, loss, and needs estimates. A 2-day full- scale exercise was conducted by SCEMD to validate the plan. An earthquake similar in size and location to the 1886 Charleston earthquake was used in the scenario.

Tennessee

The Tennessee General Assembly established a West Tennessee Seismic Safety Commission to work on policy and public preparedness issues affecting counties in the NMSZ. The Tennessee Emergency Management Agency (TEMA), CUSEC, and the University of Memphis Center for Earthquake Research and Information (CERI) are providing leadership, strategic guidance, and staff assistance to the Commission (http://itmattersareyouprepared.org).

Tennessee has completed its catastrophic disaster plan for earthquakes for 21 counties that are threatened by the NMSZ. In addition to the regional plan, each county has produced a mirror plan along with emergency support functions annexes for the county. The plans were developed with CUSEC and CERI.

The TEMA Director guided TEMA in the execution of the largest earthquake exercise in Tennessee history. In June, over 2,000 National Guard and local responders joined with dozens of federal departments and agencies in the week-long TNCAT07 earthquake exercise. Fifty-eight state agencies coordinated with over 60 local jurisdictions in testing interoperability, medical surge, hazardous materials incidents, logistics, and local CAT earthquake plans. The Tennessee Department of Transportation exercised its 1,156-page Transportation Emergency Preparedness Plan, which was awarded the "Team of Excellence" award by the Governor and the Commissioner of Transportation.

Utah

The Utah State Seismic Commission (USSC) is working on its Unreinforced Masonry Initiative with the Structural Engineers Association of Utah. The USSC has drafted a resolution to the state legislature on the need for an impact study on the economy and life safety. The USSC also is supporting a Utah State Office of Education initiative that all K-12 schools have an ATC-21 evaluation completed by a professional engineer or architect.

The Utah Geological Survey (UGS) sponsors three Utah earthquake working groups each spring that bring together more than 50 geologists, seismologists, and engineers. The Ground Shaking Working Group's goal for 2007 is to collect shear/wave data for Weber, Davis, Salt Lake, and Utah Counties; to perform deep basin model simulations and evaluate their validity; and to form a working group to develop a near/surface site amplification model. The Quaternary Fault Parameters Working Group will study three fault zones. The Liquefaction Working Group will develop permanent ground deformation maps for Salt Lake, Davis, and Utah Counties.

The University of Utah Seismograph Stations (UUSS) is expanding its network in southwestern Utah, which has had significant population increases. Central and eastern Utah will receive new instrumentation to enable the UUSS to generate ShakeMaps statewide. UGS, UUSS, USSC, the Utah Division of Homeland Security, and FEMA formed a committee to focus on generating maps, using ShakeMap and HAZUS, for emergency managers, first responders, government jurisdictions, and the private sector in the first hours after an earthquake. With support from the USGS, Utah also is developing a booklet on earthquake/related geologic hazards, the monitoring of earthquakes in Utah, how buildings perform in an earthquake, and how to plan and prepare for an earthquake. The booklet will be modeled after the USGS's Putting Down Roots in Earthquake Country.

Washington
Earthquake education materials are sent out semiannually to jurisdictions, state agencies, schools, businesses, and the general public. This is followed up with a statewide earthquake "Drop, Cover, and Hold" drill and a coastal "Tsunami Evacuation" drill. The effectiveness of public education tools is measured with social science tools. For example, a business tool kit has been developed for hotels and motels and an outreach program established to train businesses on preparedness and mitigation issues. The lack of broadcaster community understanding of state and local earthquake and tsunami procedures was also identified and a broadcaster's handbook was developed.

Washington increased its capability of alert and notification by installing over 36 all/hazard alert broadcasting radios in communities at risk for tsunamis. Special messaging was developed to warn and instruct the public on actions to take for a tsunami. Through state funding, Washington also has increased its capabilities to provide real/time earthquake information with the use of ShakeMap. This tool is being integrated into HAZUS for education and training, mitigation planning, and grant development purposes.

APPENDIX A. COOPERATING ORGANIZATIONS RECEIVING NEHRP SUPPORT

During 2007, the National Earthquake Hazards Reduction Program (NEHRP) provided partial support for the following organizations, either directly or through a second party, to advance NEHRP goals and objectives. Through participating in these cooperative efforts, NEHRP benefits from the support provided to these organizations by other interests. This list includes only those organizations cited in this report. This list does not include the many academic institutions to which NEHRP provides support for individual research grants and cooperative agreements. For each organization that is presented, a link to its Internet Web site is provided.

Applied Technology Council

The Applied Technology Council (ATC) is a nonprofit corporation established in 1973 through the efforts of the Structural Engineers Association of California. ATC's mission is to develop and promote state/of/the/art, user/friendly engineering resources and applications for use in mitigating the effects of natural and other hazards on the built environment. ATC also identifies and encourages needed research and develops consensus opinions on structural engineering issues in a nonproprietary format. Project work is conducted by a wide range of highly qualified consulting professionals, thus incorporating the experience of many individuals from academia, research, and professional practice who would not be available from any single organization. Funding for ATC projects is obtained from government agencies and from the private sector. (www.atcouncil.org)

Cascadia Region Earthquake Workgroup

The Cascadia Region Earthquake Workgroup (CREW) is a coalition of private and public representatives working together to increase the ability of Cascadia Region communities in British Columbia, California, Oregon, and Washington to reduce the effects of earthquake events. Established in 1996, CREW provides an essential link among the Federal Government, local government, private industry, and citizens to promote NEHRP goals. (www.crew.org)

Central United States Earthquake Consortium

The Central United States Earthquake Consortium (CUSEC) is a partnership of the Federal Government and the States of Alabama, Arkansas, Illinois, Indiana, Kentucky, Mississippi, Missouri, and Tennessee, the states most affected by earthquakes in the New Madrid Seismic Zone. Established in 1983, the mission of CUSEC is to reduce deaths, injuries, property damage, and economic losses resulting from earthquakes in the Central United States. (www.cusec.org)

Consortium of Organizations for Strong Motion Observation Systems

The purposes of the Consortium of Organizations for Strong Motion Observation Systems (COSMOS) are to develop policies for the urgent improvement of strong/motion earthquake measurements and their applications; promote the advancement of strong/motion measurement in densely urbanized areas and other locations of special significance to society likely to be struck by future earthquakes; encourage and assist the rapid, convenient, and responsive distribution of strong ground/motion data; and, serve as a central organization for solving problems involving the technical aspects of the acquisition and application of data on strong shaking recorded during earthquakes. (www.cosmos-eq.org)

Consortium of Universities for Research in Earthquake Engineering

The Consortium of Universities for Research in Earthquake Engineering (CUREE) is a nonprofit organization, established in 1988, which is devoted to the advancement of earthquake engineering research, education, and implementation. CUREE's membership, comprising some two dozen universities and many associated faculty members, works to identify new ways that research can solve earthquake problems; collect and synthesize information and make it easily accessible; establish national and international hazard research relationships; perform earthquake engineering and related research; manage research consortia and cooperative programs; and educate experts, practitioners, students, and the public. (www.curee.org)

Earthquake Engineering Research Institute

The Earthquake Engineering Research Institute (EERI) is a national, nonprofit technical society of engineers, geoscientists, architects, planners, public officials, and social scientists. The objective of EERI is to reduce earthquake risk by advancing the science and practice of earthquake engineering; to improve understanding of the impact of earthquakes on the physical, social, economic, political, and cultural environment; and to advocate comprehensive and realistic measures for reducing the harmful effects of earthquakes. (www.eeri.org)

Incorporated Research Institutions for Seismology

The Incorporated Research Institutions for Seismology (IRIS) is a National Science Foundation (NSF) supported university research consortium dedicated to exploring the Earth's interior through the collection and distribution of seismographic data. In addition to partnering with the U.S. Geological Survey to operate the Global Seismographic Network (GSN), NSF funding for IRIS supports the Program for Array Seismic Studies of the Continental Lithosphere (PASSCAL), which loans seismic sensors and data acquisition, telemetry, and power systems for earth science research; the IRIS Data Management System, which collects, assesses, archives, and distributes all data from the GSN, PASSCAL experiments, the Advanced National Seismic System, and other national and international sources; and the IRIS Education and Outreach Program, which enables audiences beyond seismologists to access and use seismological data and research for educational purposes. (www.iris.edu)

Multidisciplinary Center for Earthquake Engineering Research-Earthquake Engineering to Extreme Events

The Multidisciplinary Center for Earthquake Engineering Research (MCEER)-Earthquake Engineering to Extreme Events consortium is centered at the University at Buffalo, The State University of New York. Funded primarily by the NSF, the State of New York, and the Federal Highway Administration, MCEER accomplishes its mission through a system of multidisciplinary, multihazard research, education, and outreach initiatives. The goal of MCEER is to enhance the earthquake resiliency of communities through improved engineering and management tools for critical infrastructure systems (water supply, electric power, and hospitals) and emergency management functions. At the end of fiscal year (FY) 2007, MCEER completed the last year of its 10/year funding as a center supported by the NSF, but will continue as a center through state and private/sector support and other federal funding. (http://mceer.buffalo.edu)

Mid-America Earthquake Center

The Mid/America Earthquake (MAE) Center, headquartered at the University of Illinois at Urbana/Champaign, is a consortium of eight core institutions. At the end of FY 2007, the MAE Center completed the last year of its 10/year funding as a center supported by the NSF, but will continue through university and private/sector support and other federal funding. (http://mae.ce.uiuc.edu)

National Institute of Building Sciences

Congress chartered the National Institute of Building Sciences (NIBS) in 1974 as an independent, nongovernment, nonprofit organization. NIBS balances public and private expertise to mobilize uniquely authoritative support for the public interest in building sciences, engineering, construction, and technology. NIBS involves the national building community in shaping its program and priorities through its Consultative Council; other councils address specific issues in security and disaster preparedness, facility performance and sustainability, and information resources and technologies. (www.nibs.org)

Since 1979, the Building Seismic Safety Council (BSSC) of NIBS has provided a national forum for improving earthquake/resistant design and construction, benefiting both the building community and the public in general. Supported by some 65 voting member organizations, the BSSC is involved in developing the 2009 NEHRP Recommended Provisions or Seismic Regulations or New Buildings and 0ther Structures, and in working with the Federal Emergency Management Agency (FEMA) on practical building code applications of these provisions. (www.bssconline.org)

Natural Hazards Center

The NSF/supported Natural Hazards Center (NHC), headquartered at the University of Colorado at Boulder, continues to be the world leader in the dissemination of research information, awards, findings, and applications to the hazard and disaster research and management communities. The NHC accomplishes its work through four major activities: information dissemination, annual workshops, information services, and research. The majority of the Center's work is supported by an NSF grant, but the NHC also receives contributions from other agencies and sources. (www.colorado.edu/hazards/)

EES Consortium, Inc.

NEES Consortium, Inc. (NEESinc), is located in Davis, California. NEESinc is a nonprofit organization that works in partnership with the 15 universities that operate the NEES experimental facilities and cyberinfrastructure. NEESinc manages the George E. Brown, Jr. Network for Earthquake Engineering Simulation (NEES) as a national, shared/use resource for research and education for the earthquake engineering community and schedules access to the experimental facilities. NEESinc also provides the system/wide information technology infrastructure of NEES, including repositories for NEES data and simulation tools; manages an education, outreach, and training program; and fosters linkages and partnerships with federal, state, and local government entities, national laboratories, the private sector, and international collaborators. (www.nees.org)

Northeast States Emergency Consortium

The Northeast States Emergency Consortium (NESEC) receives significant funding from FEMA to support the common mission of working with federal, state, and local partners to promote multihazard preparedness and risk reduction in support of NEHRP goals. Connecticut, Maine, Massachusetts, New Hampshire, New Jersey, New York, Rhode Island, and Vermont form NESEC. (www.nesec.org)

Pacific Earthquake Engineering Research Center

Headquartered at the University of California at Berkeley, the Pacific Earthquake Engineering Research (PEER) Center focuses on areas west of the Rocky Mountains and emphasizes performance/based design in its research programs. The PEER Center's mission is to develop and disseminate procedures and supporting tools and data for performance/based earthquake engineering. This approach is aimed at improving decision/making about seismic risk by making the choice of performance goals and the tradeoffs that they entail apparent to facility owners and society at large. At the end of FY 2007, the PEER Center completed its last year of its 10/year support by the NSF; however, PEER will continue as a research center through support provided from California, the private sector, and other federal grants. (http://peer.berkeley.edu)

Southern California Earthquake Center

In 2007, the Southern California Earthquake Center (SCEC) began its third phase, a 5/year program supported primarily by the NSF and the USGS. SCEC is headquartered at the University of Southern California and unites 15 core institutions and 39 participating institutions in a "collaboratory" with a threefold mission: (1) gather data on earthquakes in southern California; (2) integrate these data and other information into a comprehensive, physics/based understanding of earthquake phenomena; and (3) communicate this understanding to the community at large as useful knowledge for reducing earthquake risk. In addition to core funding in 2007, the NSF provided support to SCEC to advance seismic hazard research using high/performance computing, with the aim of utilizing peta-scale computing facilities when they become available in the 2010– 2011 timeframe. (www.scec.org)

Western States Seismic Policy Council

The Western States Seismic Policy Council (WSSPC) is a regional earthquake consortium funded primarily by FEMA and the USGS. WSSPC members are the state Geological Surveys and Emergency Management Directors of 13 western states (Alaska, Arizona, California, Colorado, Hawaii, Idaho, Montana, Nevada, New Mexico, Oregon, Utah, Washington, and Wyoming); 3 U.S. territories (American Samoa, Guam, and the Northern Mariana Islands); a Canadian territory (Yukon Territory); and a Canadian province (British Columbia). The mission of the WSSPC is to develop seismic policies and share information to promote programs intended to reduce earthquake losses. (www.wsspc.org)

B. EHRP MANAGEMENT CHRONOLOGY FY 2007

Table B.1 provides a brief chronology of National Earthquake Hazards Reduction Program (NEHRP) management activities during fiscal year (FY) 2007.

ACEHR—Advisory Committee on Earthquake Hazards Reduction

The members are nongovernment experts. ACEHR was created by statute, P.L. 108/360.

ICC—Interagency Coordinating Committee on Earthquake Hazards Reduction

The principals are the Directors/Administrators of the NEHRP agencies and of the Office of Science and Technology Policy and Office of Management and Budget. The ICC was created by statute, P.L. 108/360.

PCWG—Program Coordination Working Group

A working/level body made up of the relevant program managers of each NEHRP agency.

Table B.1—Chronology of FY 2007 NEHRP Management Activities

Date	Event
October 3, 2006	ICC Meeting
November 16, 2006	PCWG Meeting
December 15, 2006	PCWG Meeting
January 25, 2007	PCWG Meeting
February 16, 2007	PCWG Meeting
March 22, 2007	PCWG Meeting
May 10–11, 2007	First Meeting of ACEHR
May 21, 2007	PCWG Meeting
May 30, 2007	NEHRP Report for FY 2005–2006 Submitted to Congress
May 31, 2007	ICC Meeting
June 26, 2007	PCWG Meeting
July 23, 2007	PCWG Meeting
August 27, 2007	PCWG Meeting
September 10, 2007	PCWG Meeting
September 25, 2007	ICC Meeting

C. List of Acronyms

ACCESS	Advancement of Cyberinfrastructure Careers through Earthquake System Science
ACEHR	Advisory Committee on Earthquake Hazards Reduction
ACI	American Competitiveness Initiative
AEIC	Alaska Earthquake Information Center
AMP	Amplitude and Period Monitoring
ANSS	Advanced National Seismic System
ASCE	American Society of Civil Engineers
ATC	Applied Technology Council
BMG	Indonesia Meteorological and Geophysical Agency
BSSC	Building Seismic Safety Council
CCERRP	California Catastrophic Earthquake Readiness Response Plan
CDMS	Comprehensive Data Management System
CERI	Center for Earthquake Research and Information
CGS	California Geological Survey
CI	Cyberinfrastructure
CISN	California Integrated Seismic Network
CONPLAN	Concept of Operations Plan
COSMOS	Consortium of Strong Motion Observation Systems
CREW	Cascadia Region Earthquake Workgroup
CRM	Consequence/based Risk Management
CUSEC	Central United States Earthquake Consortium
DMS	IRIS Data Management System
DOGAMI	Oregon Department of Geology and Mineral Industries
DOI	Department of the Interior
DPI	Community Disaster Preparedness Index
DRI	Community Disaster Response Index
EAS	Emergency Alert System
ECA	Earthquake Country Alliance
EERI	Earthquake Engineering Research Institute
EEW	Earthquake Early Warning
ETS	Episodic Tremor and Slip
FEMA	Federal Emergency Management Agency
FY	Fiscal Year
GEOSS	Global Earth Observation System of Systems
GIS	Geographic Information System
GPS	Global Positioning System; Global Positioning Satellites

GSN	Global Seismographic Network
HAZUS	Hazards U.S.
HDPE	High/density polyethylene
HISN	Hawaii Integrated Seismic Network
HMGP	Hazard Mitigation Grant Program
IBC	International Building Code
ICC	International Code Council; Interagency Coordinating Committee
IGS	Idaho Geological Survey
IQR	Internet Quick Reports
IRC	International Residential Code
IRIS	Incorporated Research Institutions for Seismology
IT	Information Technology
KBL	Kabul Seismic Station
LiDAR	Light Detection and Ranging
MAE	Center Mid/America Earthquake Center
MCEER	Multidisciplinary Center for Earthquake Engineering Research
NCESMD	National Center for Engineering Strong Motion Data
NCEDC	Northern California Earthquake Data Center
NEES	George E. Brown, Jr. Network for Earthquake Engineering Simulation
NEHRP	National Earthquake Hazards Reduction Program
NEIC	National Earthquake Information Center
NEMRAC	NESEC Emergency Management Risk Assessment Center
NEPEC	National Earthquake Prediction Evaluation Council
NESC	Nevada Earthquake Safety Council
NESEC	Northeast States Emergency Consortium
NETAP	National Earthquake Technical Assistance Program
NGDC	National Geophysical Data Center
NHC	Natural Hazards Center
NIBS	National Institute of Building Sciences
NIED	National Research Institute for Earth Science and Disaster Prevention
NIST	National Institute of Standards and Technology
NMSZ	New Madrid Seismic Zone
NOAA	National Oceanic and Atmospheric Administration
NRC	National Research Council
NSF	National Science Foundation
NSMP	National Strong Motion Project

NSTC	National Science and Technology Council
OEM	Oregon Emergency Management
PACT	Performance Assessment Calculation Tool
PAGER	Prompt Assessment of Global Earthquakes for Response
PASSCAL	Program for Array Seismic Studies of the Continental Lithosphere
PBO	Plate Boundary Observatory
PBSD	Performance/Based Seismic Design
PCWG	Program Coordination Working Group
PDM	Pre/Disaster Mitigation Grant
PDM-C	Pre/Disaster Mitigation/Competitive Grant
PEER	Pacific Earthquake Engineering Research Center
PG&E	Pacific Gas & Electric
PIMS	Post/Disaster Information Management System
PNSN	Pacific Northwest Seismic Network
PREMA	Puerto Rico Emergency Management Agency
PTWC	Pacific Tsunami Warning Center
RVS	Rapid Visual Screening
SAFOD	San Andreas Fault Observatory at Depth
SCEC	Southern California Earthquake Center
SCEMD	South Carolina Emergency Management Division
SCSN	Southern California Seismic Network
SDR	Subcommittee on Disaster Reduction
SONS	Spills of National Significance
SoSAFE	Southern San Andreas Fault Evaluation
TA	Transportable Array
TDOT	Tennessee Department of Transportation
TEMA	Tennessee Emergency Management Agency
UAF/GI	University of Alaska/Geophysical Institute
UC	University of California
UGS	Utah Geological Survey
UJNR	U.S./Japan Program on Natural Resources
UNR	University of Nevada, Reno
USGS	U.S. Geological Survey
USSC	Utah Seismic Safety Commission
UUSS	University of Utah Seismic Stations
VDC	Virtual Data Center
VS	Virtual Seismologist
WSDOT	Washington State Department of Transportation

WSSPC   Western States Seismic Policy Council

In: Earthquakes: Risk, Monitoring and Research ISBN: 978-1-60692-648-2
Editor: Earl V. Leary © 2009 Nova Science Publishers, Inc.

Chapter 3

FORECASTING CALIFORNIA'S EARTHQUAKES—WHAT CAN WE EXPECT IN THE NEXT 30 YEARS?

In a new comprehensive study, scientists have determined that the chance of having one or more magnitude 6.7 or larger earthquakes in the California area over the next 30 years is greater than 99%. Such quakes can be deadly, as shown by the 1989 magnitude 6.9 Loma Prieta and the 1994 magnitude 6.7 Northridge earthquakes. The likelihood of at least one even more powerful quake of magnitude 7.5 or greater in the next 30 years is 46%—such a quake is most likely to occur in the southern half of the State. Building codes, earthquake insurance, and emergency planning will be affected by these new results, which highlight the urgency to prepare now for the powerful quakes that are inevitable in California's future.

WHAT IS AN EARTHQUAKE RUPTURE FORECAST?

Californians know that their State is subject to frequent—and sometimes very destructive—earthquakes. Accurate forecasts of the likelihood of quakes can help people prepare for these inevitable events. Because scientists cannot yet make precise predictions of the date, time, and place of future quakes, forecasts are in the form of the probabilities that quakes of certain sizes will occur during specified periods of time.

In our daily lives, we are used to making decisions based on probabilities—from weather forecasts (such as a 30% chance of rain) to the annual chance of being killed by lightning (about 0.0003%). Similarly, earthquake probabilities derived by scientists can help us plan and prepare for future quakes.

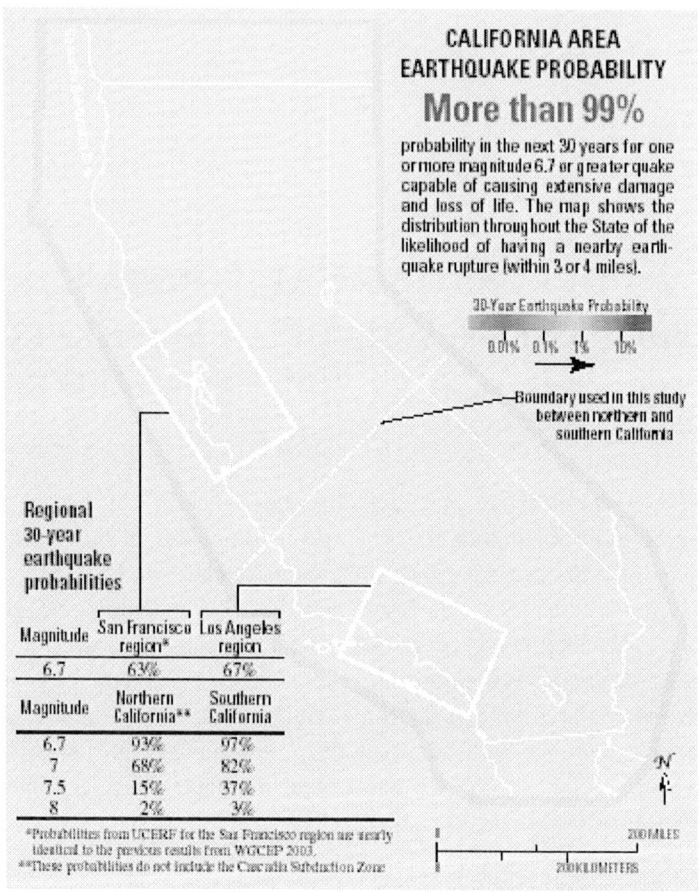

Earthquake forecasts for California have been developed in the past by multidisciplinary groups of scientists and engineers, each known as a "Working Group on California Earthquake Probabilities" (WGCEP 1988, 1990, 1995, 2003). However, those forecasts were limited to particular regions of California. Because of this, WGCEP 2007 was commissioned to develop an updated, statewide forecast, the latest result of which is the Uniform California Earthquake Rupture Forecast, Version 2, or "UCERF" (U.S. Geological Survey (USGS) Open-File Report 2007-1437, http://pubs.usgs. gov/of/2007/1437/). Organizations sponsoring WGCEP 2007 include the USGS, California Geological Survey, and the Southern California Earthquake Center. The comprehensive new forecast builds on previous studies and also incorporates abundant new data and improved scientific understanding of earthquakes. When an earthquake occurs, two things happen—a fault ruptures (a crack in the Earth's crust gives way and slips under tectonic pressure) and seismic waves, caused by this sudden fault motion, radiate out like ripples from a pebble tossed into a pond. The shaking that occurs as seismic waves pass by causes most quake damage. The strength of the waves depends partly on the quake's magnitude, which is a function of the size of the fault that moves and the amount of slip.

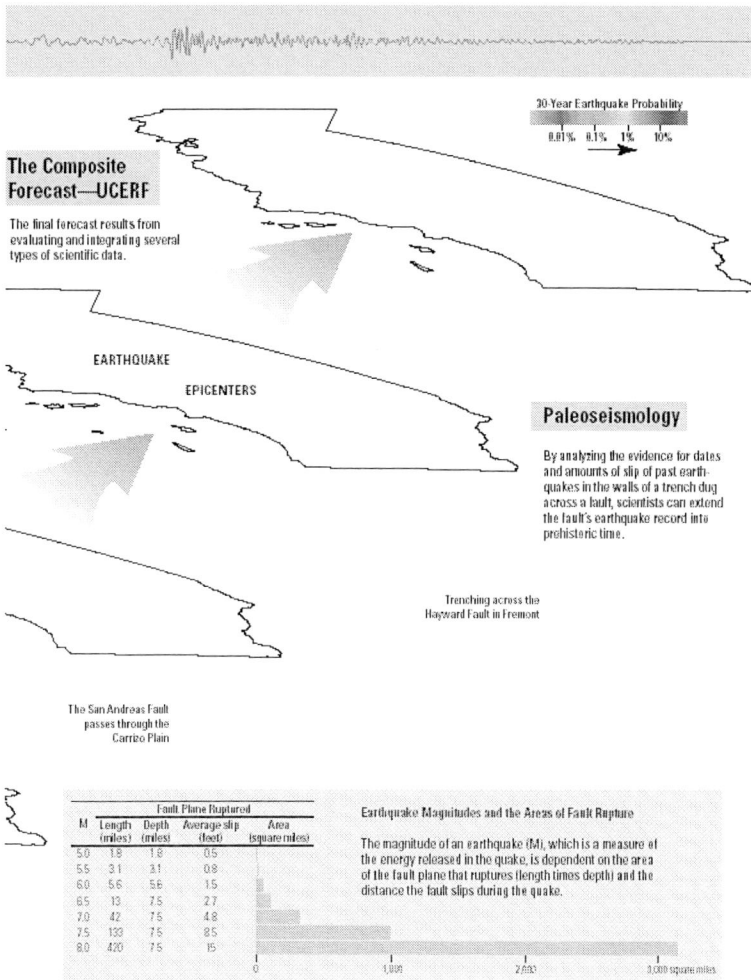

The UCERF study's goal was to determine probabilities for different parts of California of earthquake ruptures of various magnitudes, but not to estimate the likelihood of shaking ("seismic hazard") that will be caused by these quakes. This distinction is important, because even areas in the State with a low probability of fault rupture can experience shaking and damage from distant, powerful quakes.

How Likely is a Damaging Quake in the Next 30 Years?

California straddles the boundary between two of the Earth's tectonic plates—as a result, it is broken by numerous earthquake faults. Taking into account the earthquake histories and relative rates of motion on these many faults, the UCERF study concludes that there is a probability of more than 99% that in the next 30 years Californians will experience one or more magnitude 6.7 or greater quakes, potentially capable of causing extensive damage and loss of life. For powerful quakes of magnitude 7.5 or greater, there is a 46% chance of one or more in the next 30 years—such a quake is twice as likely to occur (37%) in the southern half of the State than in the northern half (15%).

Smaller magnitude earthquakes are more frequent than larger quakes. According to the new forecast, about 3 magnitude 5 or greater quakes will occur in the California region per year, and a magnitude 6 or greater quake about every 1.5 years. These numbers do not include aftershocks that follow larger quakes—including them would roughly double the expected number of magnitude 5 or greater quakes.

STATEWIDE EARTHQUAKE PROBABILITIES

The numbers represent current best estimates. As earthquake science progresses, these probabilities will change. Actual repeat times vary considerably and only rarely will be exactly as listed in the table.

Magnitude	30-year probability of one or more events greater than or equal to the magnitude	Average repeat time (years)
6.7	>99%	5
7	94%	11
7.5	46%	48
8	4%	650

*Not including Cascadia Subduction Zone

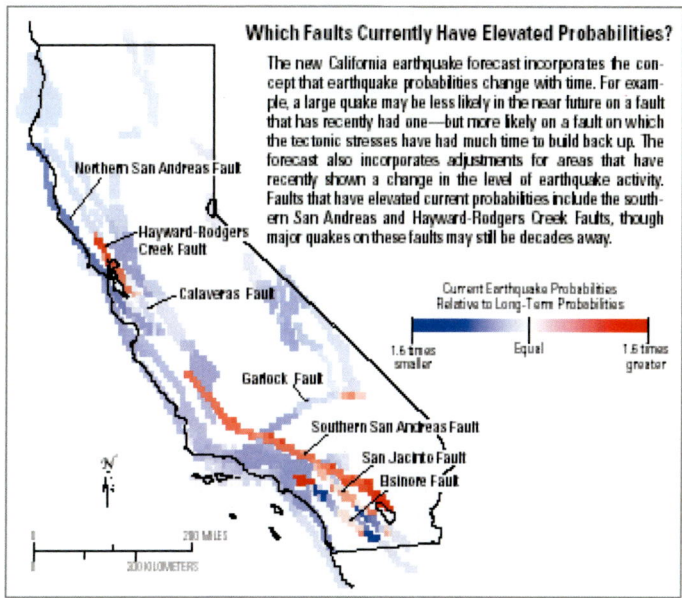

For the entire California region, the fault with the highest probability of generating at least one magnitude 6.7 or larger earthquake is the southern San Andreas (59% in the next 30 years). For northern California, the most likely source of such a quake is the Hayward-Rodgers Creek Fault (31% in next 30 years)—see USGS Fact Sheet 2008-3019. Quake probabilities for many parts of the State are similar to those in previous studies, but the new probabilities for the Elsinore and San Jacinto Faults in southern California are about half those previously determined. For the far northwestern part of the State, a major source of quakes is the offshore 750-mile-long "Cascadia Subduction Zone," which extends south about 150 miles into California. For the next 30 years there is a 10% probability of a magnitude 8 to 9 quake somewhere along the zone—such quakes occur about every 500 years.

The UCERF forecast was evaluated by an independent scientific review panel, as well as by both the California and National Earthquake Prediction Evaluation Councils, making it one of the most extensively reviewed earthquake forecasts ever produced. Uncertainties remain because the new quake probabilities are the result of evaluating and accommodating several earthquake theories. As scientific understanding of quakes improves, the probabilities will change. The results of the UCERF study are a reminder that all Californians live in earthquake country and should therefore be prepared (see Putting Down Roots in Earthquake Country at http://www.earthquakecountry.info/roots/). The USGS has already used the UCERF to estimate California's seismic hazard, which in turn will be used to update building codes. Other subsequent studies will add information on the vulnerability of manmade structures to estimate expected losses ("seismic risk"). In these ways, UCERF will help to increase public safety and community resilience to earthquake hazards. Earthquakes cannot be prevented, but the damage they do can be greatly reduced through prudent planning and preparedness. The ongoing work of USGS, California Geological Survey, Southern California Earthquake Center, and other scientists in evaluating quake probabilities is part of the National Earthquake Hazard Reduction Program's efforts to safeguard lives and property from the future quakes that are certain to strike in California and elsewhere in our Nation.

INDIVIDUAL FAULT PROBABILITIES

The UCERF report assigns individual probabilities to specific known major faults. Below are 30-year probabilities for seven of the faults for which scientists have the most data. Many other faults also have significant probabilities—in fact, the next big quake in California is just as likely to occur on one of the other faults in the State.

Fault	Probability of one or more magnitude 6.7 or greater quake
Southern San Andreas	59%
Hayward-Rodgers Creek	31%
San Jacinto	31%
Northern San Andreas	21%
Elsinore	11%
Calaveras	7%
Garlock	6%

INDEX

A

academic, 22, 33, 49, 51, 55, 74, 75, 79, 83, 100
academics, 33
access, 33, 52, 58, 79, 90, 103, 105
accounting, 10
accuracy, 73
ACI, 108
acute, 37, 56, 97
administrators, 89
advocacy, 68
Afghanistan, 89, 90
aging, 13, 30
aid, 20, 38, 46, 62
air, 45
Alabama, 63, 101
Alaska, vii, 1, 3, 6, 13, 17, 34, 48, 75, 91, 93, 106, 108, 110
amendments, 64
American Competitiveness Initiative, 36, 108
amplitude, 22
analog, 76
application, 20, 46, 102
applied research, 30, 34, 53, 88
appropriations, viii, 2, 3, 15, 20
Appropriations Committee, 36
Arizona, 86, 106
Arkansas, 63, 70, 101
arteries, 61

Asia, 90
Asian, 68, 92
assessment, 38, 46, 47, 51, 53, 63, 78, 91, 93, 94, 95
ATC, 53, 55, 67, 93, 95, 97, 98, 99, 101, 108
Atlantic, 91
availability, 90
averaging, 10
awareness, 15, 47, 68, 70, 94, 97, 98

B

bandwidth, 82
Bangladesh, 8
Barbados, 82
barriers, 88
basic research, 34, 53
behavior, 9, 15, 22, 37, 61, 81, 85
Beijing, 8, 89
benefits, viii, 2, 12, 24, 88, 100
blocks, 42
board members, 70
Boston, 6, 11, 12, 27, 78
branching, 85
Brazil, 95
British Columbia, 101, 106
broadband, 13, 14, 75, 78, 79, 89, 90
broadcaster, 100
bubbles, 45
buffalo, 103

buildings, vii, 1, 4, 5, 8, 9, 13, 14, 17, 19, 20, 21, 23, 32, 36, 37, 40, 42, 50, 52, 53, 54, 56, 58, 66, 80, 88, 90, 99
Bureau of Land Management, 51

C

California Earthquake Authority, 68, 70
Canada, 27, 86
capacity, 91, 92
capacity building, 91
Caribbean, 14, 82, 91, 92, 97
CAT, 99
catastrophes, 24
CEC, 44
centralized, 83
channels, 16, 76, 79
children, 98
Chile, 48, 82
China, vii, 1, 2, 8, 88, 89
Christmas, 82
citizens, 3, 6, 93, 98, 101
civil engineering, 87
classes, 95
clusters, 27
Co, 88
Coast Guard, 82
coastal communities, 64
codes, viii, 12, 20, 32, 33, 36, 45, 53, 64, 65, 66, 67, 87, 92, 113, 118
collaboration, 32, 38, 49, 61, 67, 90
Colorado, 28, 34, 66, 73, 86, 96, 104, 106
Columbia, 78
Columbia University, 78
communication, 17, 27, 47, 62, 81
communities, 33, 36, 40, 41, 42, 46, 47, 48, 52, 64, 66, 68, 70, 88, 93, 94, 100, 101, 103, 104
community, 47, 48, 52, 53, 61, 63, 69, 79, 100, 104, 105, 106, 118
Community Development Block Grant, 24
competitiveness, 30
complexity, 60
compliance, 14
components, 10, 14, 15, 47, 54, 56, 57, 63, 83

composition, 83, 84
Comprehensive Test Ban Treaty, 14, 82
computation, 17
computer simulations, 23, 51
computing, 106
conception, 13
concrete, 37, 57, 65, 81, 88
Congress, viii, 2, 3, 12, 13, 19, 20, 24, 25, 28, 30, 32, 33, 35, 39, 40, 64, 78, 83, 104, 107
Congressional Budget Office, 10
Connecticut, 105
consensus, 42, 63, 65, 101
construction, 10, 12, 20, 28, 30, 31, 33, 45, 53, 66, 79, 86, 87, 89, 95, 104
consulting, 101
continuity, 78
contractors, 32, 36
contracts, 28
convergence, 8
Cook Islands, 82
cost-effective, 30, 40, 41
costs, viii, 2, 3, 12, 21
coverage, 13, 43, 90
covering, 13, 37, 88
crack, 115
creep, 61, 71, 84, 85
critical infrastructure, 8, 19, 48, 103
CRM, 108
CRS, 1, 4, 24, 25, 27
crust, 8, 22, 23, 42, 83, 89, 115
curriculum, 96

D

data analysis, 32, 34, 38, 73, 78
data availability, 91
data collection, 48
data set, 52, 76, 83
database, 75, 88, 91, 97
deaths, 6, 7, 26, 60, 70, 101
decisions, 12, 89, 113
deformation, 22, 44, 61, 83, 85, 86, 88, 99
Delaware, 47
delivery, 63, 73
demand, 67

demographics, 30
Demonstration Project, 38, 51
density, 3, 13, 48, 56, 109
Department of Commerce, 16
Department of Homeland Security, 25, 41
Department of the Interior, 74, 108
Department of Transportation, 58, 64, 79, 97, 99, 110
deposits, 45
destruction, 25, 91
digital communication, 17
disaster, 21, 22, 32, 33, 36, 38, 46, 48, 52, 59, 62, 64, 69, 73, 78, 80, 87, 88, 90, 93, 94, 95, 98, 104
disseminate, 20, 28, 105
distribution, 13, 47, 51, 52, 58, 60, 62, 94, 95, 102, 103
Dominican Republic, 82
donations, 47

E

early warning, 60
earth, 22, 34, 43, 44, 83, 86, 95, 103
Earth Science, 23, 88, 109
economic activity, 9
economic losses, vii, 1, 2, 3, 8, 10, 101
Ecuador, 82
Education, 72, 93, 98, 99, 103
educational programs, 81
educators, 95
electric power, 103
elementary teachers, 95
email, 16
emergency management, 52, 61, 69, 93, 96, 103
emergency planning, ix, 113
emergency preparedness, 47
emergency response, 13, 14, 15, 16, 17, 22, 51
Emergency Supplemental Appropriations Act, 14
employees, 70
energy, 9, 25, 44, 86, 89, 97
enterprise, 87

environment, 9, 28, 37, 40, 41, 48, 53, 59, 80, 92, 101, 102
estimating, 17, 45
Ethiopia, 82
Eureka, 11
evolution, 23, 83
execution, 34, 99
exercise, 38, 61, 62, 63, 94, 96, 97, 98, 99
expenditures, 14
expert, 31
expertise, 33, 35, 44, 47, 87, 90, 92, 104
explosions, 13, 14
exposure, 9, 32, 38, 59

F

faculty positions, 72
failure, 47, 71
fatalities, 2, 6, 7, 8, 26, 89, 91
faults, 9, 22, 26, 42, 43, 44, 49, 61, 86, 89, 97, 116
fax, 16
February, 7, 11, 26, 57, 68, 72, 81, 86, 96, 107
Federal Emergency Management Agency, viii, 2, 25, 31, 33, 36, 45, 93, 104, 108
Federal Emergency Management Agency (FEMA), viii, 2, 31, 33, 45, 93, 104
federal government, 2
federal grants, 105
Federal Highway Administration, 103
feet, 57, 84
FEMA, v, viii, 2, 3, 11, 19, 20, 21, 23, 25, 26, 27, 29, 31, 32, 33, 36, 41, 45, 51, 52, 53, 54, 55, 56, 63, 64, 65, 66, 67, 68, 69, 70, 73, 93, 94, 95, 97, 98, 99, 104, 105, 106, 108
fiber, 57, 79
fire, 47, 97
fires, 6, 51, 56, 63
first responders, 99
flow, 22, 45
fluid, 22
focusing, 63
forecasting, 42
Forest Service, 51

Forestry, 43
funding, viii, 2, 15, 19, 20, 24, 31, 32, 36, 37, 78, 95, 100, 103, 104, 105, 106
funds, 23, 24, 39, 66, 78, 93, 97

G

gas, 45, 56, 94, 95
gauge, 59
generation, 37, 44, 75, 77, 84, 92
Geographic Information System, 108
geology, 22, 42, 50, 73, 77, 95, 97
geothermal, 76
geothermal field, 76
GIS, 21, 52, 70, 108
global economy, 30
Global Positioning System, 83, 108
Global War on Terror, 14
goals, 24, 31, 34, 39, 41, 83, 87, 94, 100, 101, 105
governance, 48
government, 2, 14, 16, 19, 30, 37, 40, 49, 51, 58, 63, 79, 89, 91, 93, 99, 101, 105
GPS, 23, 44, 61, 83, 86, 89, 92, 108
grants, 23, 34, 44, 93, 100, 105
gravity, 3
Grenada, 82
groups, 16, 20, 22, 26, 33, 37, 52, 62, 72, 78, 99, 114
growth, 43, 44, 89
Guam, 106
Guantanamo, 82
guidance, 54, 93, 98
guidelines, 45, 53, 54, 80, 88
Gulf Coast, 46

H

handling, 47
hands, 95
harmful effects, 102
Hawaii, vii, 1, 3, 13, 48, 67, 77, 91, 94, 95, 106, 109
hazardous materials, 99

hazards, vii, viii, 1, 2, 3, 4, 9, 10, 11, 13, 15, 19, 21, 22, 29, 40, 44, 47, 48, 49, 51, 52, 53, 61, 63, 64, 66, 68, 70, 86, 87, 89, 91, 93, 97, 99, 101, 104, 118
HDPE, 56, 57, 109
hearing, 88
heat, 22
height, 91
high school, 95
high-risk, 13
highways, 8
Homeland Security, 25, 41, 78, 93, 95, 99
Honduras, 82
hospitals, 36, 56, 67, 90, 103
host, 44, 84
hotels, 100
House, 20, 28, 33, 36
households, 21, 69
housing, 70
human, 17, 71
hurricane, 21, 82
Hurricane Katrina, vii, 1, 9, 46, 69
hurricanes, 9, 11, 15, 26
hybrid, 31, 37, 46

I

IDA, 15
Idaho, 6, 78, 86, 95, 106, 109
Illinois, 28, 46, 63, 95, 101, 103
imagery, 43
images, 4
imaging, 86
imaging techniques, 86
impact assessment, 46
implementation, 20, 30, 33, 34, 35, 36, 39, 42, 57, 67, 72, 74, 77, 102
in situ, 47
inclusion, 38, 49, 64
income, 10
Indian, 48, 59, 91, 92
Indian Ocean, 48, 91, 92
Indiana, 6, 63, 70, 95, 101
indices, 47
Indonesia, 48, 92, 108

Index

industry, 49, 51, 56, 79, 81, 96, 101
information technology, 74, 79, 105
Information Technology, 109
infrastructure, vii, 1, 3, 4, 8, 9, 12, 19, 21, 28, 30, 31, 37, 46, 47, 48, 60, 79, 80, 90, 103, 105
injuries, 26, 101
institutions, 14, 15, 20, 28, 42, 49, 71, 74, 75, 79, 83, 90, 100, 103, 106
instruction, 95
instructors, 98
instruments, 13, 14, 17, 23, 25, 44, 78, 83, 84, 85
insurance, ix, 113
integration, 47, 63, 76, 91, 97
integrity, 31
intensity, 17, 27, 51
interaction, 37, 69, 80
interactions, 88
interdisciplinary, 19, 40, 42
interface, 71
Internet, 37, 75, 100, 109
interoperability, 99
interval, 70
inventories, 17
investment, 89
Iran, 48
IRC, 55, 64, 109
iris, 103
Islamic, 89
island, 94, 98
Italy, 48

J

Jamaica, 82
January, 7, 18, 73, 76, 86, 94, 107
Japan, 26, 44, 48, 56, 60, 87, 88, 110
Japanese, 87, 88, 90
jobs, 21
Jordan, 24
jurisdictions, 32, 36, 52, 78, 95, 99, 100

K

K-12, 99
Kashmir, 48
Katrina, vii, 1, 9, 46, 69
Kentucky, 63, 70, 96, 101
King, 24
Kiribati, 82
Kobe, 56
Korean, 68

L

land, 48
land use, 48
large-scale, 20, 57
law, 21, 24
lawyers, 78
lead, viii, 2, 19, 20, 23, 31, 33, 43, 56
leadership, 23, 98
learning, 95
legislation, 39, 97
lifespan, 28
likelihood, viii, 5, 42, 113, 116
limitations, 17
links, 52, 61, 70, 82
liquefaction, 45, 88
loans, 103
local authorities, 16
local government, 19, 40, 58, 79, 93, 101, 105
location, 14, 15, 16, 27, 60, 73, 76, 89, 90, 98
logistics, 99
long period, 45
long-term, viii, 2, 19
Los Angeles, vii, 1, 2, 6, 9, 10, 11, 26, 28, 38, 51, 58, 61, 62, 66, 72, 94
losses, viii, 2, 9, 10, 11, 12, 19, 21, 24, 29, 30, 31, 32, 48, 51, 106, 118
low risk, 3
lying, 9

M

magmatic, 83

Maine, 78, 105
maintenance, 39, 82
Malaysia, 92
management, 14, 20, 22, 31, 33, 35, 36, 46, 48, 52, 61, 67, 69, 77, 79, 92, 93, 96, 103, 104, 106
man-made, 22
mantle, 22, 83
mapping, 43, 44, 88, 93
Mariana Islands, 106
Maryland, 27
masonry, 37, 51, 81, 88
Massachusetts, 6, 27, 61, 105
measurement, 87, 102
measures, viii, 2, 19, 30, 34, 40, 41, 42, 47, 49, 57, 63, 88, 94, 98, 102
mechanical behavior, 22, 85
media, 10, 16, 26, 60, 68, 70, 78, 97
membership, 28, 67, 102
memorandum of understanding, 90
memory, 12
messages, 59
metropolitan area, 43
Mexican, 61
Mexico, 6, 34, 48, 60, 86, 106
military, 62
minerals, 84
mining, 78
Ministry of Education, 90
Minnesota, 28
mirror, 99
Mississippi, 12, 52, 63, 79, 101
Mississippi River, 79
Missouri, 6, 63, 67, 70, 96, 101
modeling, 26, 45, 88, 95
models, 5, 22, 42, 43, 57, 59, 63, 85, 89
momentum, 31
Montana, 86, 106
motels, 100
motion, 3, 5, 10, 12, 14, 17, 22, 25, 27, 28, 42, 44, 58, 70, 75, 76, 78, 79, 84, 89, 102, 115, 116
movement, 8, 61
multidisciplinary, 46, 103, 114

N

NASA, 23, 83
nation, 13, 30
national, 16, 19, 30, 32, 36, 38, 58, 63, 64, 65, 66, 73, 79, 81, 83, 87, 91, 92, 102, 103, 104, 105
National Bureau of Standards, 19
National Guard, 96, 99
National Institute of Standards and Technology, viii, 2, 3, 29, 31, 33, 36, 53, 87, 109
National Institute of Standards and Technology (NIST), viii, 2, 3, 31, 33
National Oceanic and Atmospheric Administration, 51, 91, 93, 109
National Oceanic and Atmospheric Administration (NOAA), 51, 91, 93
National Research Council, viii, 29, 80, 109
National Science and Technology Council, 87, 91, 110
National Science Foundation, v, viii, 2, 3, 23, 29, 31, 33, 34, 37, 42, 83, 103, 109
natural, vii, viii, 1, 2, 15, 19, 22, 24, 29, 40, 42, 48, 51, 53, 60, 66, 87, 93, 101
natural disasters, 2, 48
natural hazards, viii, 15, 22, 29, 48, 51, 66, 87, 93
natural science, 19, 40, 42
natural sciences, 19, 40, 42
network, 12, 13, 15, 27, 32, 33, 38, 47, 60, 73, 74, 76, 77, 78, 79, 81, 86, 92, 99
Nevada, 6, 17, 28, 57, 86, 97, 106, 109, 110
New Jersey, 9, 27, 105
New Mexico, 6, 34, 86, 106
New York, 6, 9, 11, 12, 26, 27, 28, 103, 105
next generation, 81
NHC, 66, 104, 109
nickel, 57
NIST, viii, 2, 19, 20, 21, 23, 31, 33, 35, 36, 40, 41, 65, 109
NOAA, 51, 59, 73, 74, 77, 82, 91, 92, 98, 109
nonlinear, 88
non-profit, 28
North America, 23, 26, 83, 84, 86

Index

North Carolina, 47
Northeast, 26, 52, 61, 70, 78, 105, 109
NRC, viii, 29, 109
nuclear, 13, 14
nucleation, 84

O

observations, 25, 44, 85
obsolete, 13, 30
Office of Management and Budget, 20, 31, 35, 107
Office of Science and Technology Policy, 20, 31, 35, 107
offshore, 5, 117
online, 52, 70
OR, 11
Oregon, vii, 1, 3, 5, 6, 10, 28, 43, 64, 68, 78, 86, 92, 97, 101, 106, 108, 110
organization, 28, 33, 53, 79, 100, 101, 102, 104, 105
organizations, 15, 16, 19, 22, 33, 34, 37, 58, 62, 67, 70, 100, 104
oversight, 17, 31, 33, 35, 71

P

PA, 11
Pacific, 17, 23, 30, 37, 43, 44, 45, 63, 68, 71, 77, 82, 84, 86, 91, 94, 97, 105, 110
Pakistan, 48
Panama, 82
partnership, 23, 31, 38, 44, 51, 61, 79, 83, 86, 92, 94, 101, 105
partnerships, 68, 74, 75, 79, 90, 105
pathways, 72
peer, 105
Pennsylvania, 27
performance, 20, 31, 36, 37, 39, 45, 48, 52, 53, 54, 55, 57, 65, 72, 73, 78, 80, 88, 90, 92, 93, 104, 105, 106
periodic, 30
personal, 27
personal communication, 27

Peru, 48
PG, 43, 68, 76, 110
Philadelphia, 9, 11, 26
Phoenix, 62
phone, 9
physical properties, 84
physics, 84, 106
pipelines, 56, 90
PL, 28
planning, ix, 19, 20, 21, 35, 36, 39, 47, 48, 52, 58, 63, 64, 67, 69, 93, 94, 95, 96, 100, 113, 118
plants, 76
plastic, 56
play, 19, 87
PM, 8
police, 97
policy makers, 66, 87
polyethylene, 56, 109
polymer, 57
polymer composites, 57
pond, 115
population, 3, 4, 6, 30, 32, 38, 52, 59, 99
pore, 45
ports, 37, 81, 94
power, 56, 90, 103
power plant, 90
power plants, 90
predictability, 42
prediction, 15, 19, 88
preparedness, 19, 47, 51, 52, 62, 68, 70, 80, 93, 98, 100, 104, 105, 118
pressure, 22, 45, 115
prevention, 90
priorities, 20, 30, 55, 63, 68, 104
private, 20, 22, 28, 30, 31, 35, 37, 44, 51, 70, 76, 79, 89, 94, 99, 101, 103, 104, 105
private sector, 28, 31, 35, 37, 44, 79, 99, 101, 105
private-sector, 20
probability, 4, 5, 10, 12, 15, 44, 50, 77, 116, 117
producers, 66
production, 75, 77
profit, 28

program, viii, 2, 14, 19, 20, 21, 22, 23, 24, 31, 34, 35, 36, 37, 42, 43, 44, 46, 49, 58, 60, 67, 72, 73, 79, 80, 81, 83, 84, 91, 92, 93, 97, 100, 104, 105, 106, 107
promote, 20, 52, 55, 78, 87, 101, 102, 105, 106
property, 9, 26, 101, 118
protocol, 69
protocols, 69, 81
prototype, 52, 79
public, 15, 16, 22, 30, 33, 47, 51, 52, 59, 60, 61, 68, 70, 71, 73, 76, 87, 88, 89, 94, 97, 98, 100, 101, 102, 104, 118
public awareness, 15, 94, 98
public domain, 76
public education, 47, 89, 100
public interest, 104
public policy, 33
public safety, 118
public schools, 97, 98
public service, 68
Puerto Rico, 6, 97, 98, 110

R

radio, 98
rain, 113
range, 2, 44, 73, 88, 101
real time, 61, 89, 91
reality, 47
reconstruction, 21, 89
recovery, 21, 24, 46, 47, 64, 69, 80
recruiting, 36
recurrence, 63, 91
Red Cross, 16, 68, 70
reduction, viii, 2, 19, 20, 30, 33, 35, 40, 42, 45, 49, 58, 63, 67, 70, 87, 88, 105
regional, 7, 13, 15, 17, 26, 27, 35, 37, 42, 50, 58, 60, 63, 64, 67, 73, 74, 75, 91, 94, 98, 106
regulation, 47
regulations, 52
regulators, 53
rehabilitation, 32, 36, 65, 67, 80, 97
relationship, 17, 47, 64, 97

relationships, 39, 42, 102
relative size, 25
reliability, 54
repair, 21
research, viii, 2, 3, 4, 13, 14, 19, 20, 22, 23, 24, 27, 28, 30, 31, 34, 35, 36, 37, 40, 42, 44, 46, 48, 49, 51, 53, 54, 55, 56, 57, 60, 61, 65, 66, 71, 72, 73, 79, 80, 83, 87, 88, 89, 90, 92, 100, 101, 102, 103, 104, 105, 106
research and development, 87
researchers, 12, 45, 51, 53, 55, 57, 61, 66, 87, 92
reservoir, 76
residential, 21
resilience, 24, 40, 41, 47, 56, 118
resistance, 8, 42, 59
resolution, 13, 43, 77, 86, 99
resources, 47, 52, 53, 70, 81, 101, 104
responsibilities, 20, 32
restructuring, 20
Rhode Island, 105
risk, vii, viii, 1, 2, 3, 4, 6, 10, 12, 13, 19, 22, 29, 30, 32, 33, 37, 42, 46, 49, 54, 57, 63, 69, 70, 73, 74, 81, 88, 91, 94, 97, 100, 102, 105, 106, 118
risk assessment, 49, 91, 97
risk management, 22, 46
risks, 3, 4, 20, 30, 68, 93
Roads, 25
rods, 57
routing, 78
RVS, 67, 95, 97, 98, 110

S

safeguard, 118
safety, 19, 20, 30, 31, 36, 37, 53, 69, 71, 93, 97, 98, 99
salary, 73
Samoa, 106
sample, 66, 85
satellite, 82
saturation, 45
school, 21, 56, 94, 95, 96, 97, 98, 99, 100

science education, 83, 95
scientific community, 69
scientific knowledge, 21
scientific understanding, 15, 22, 115, 117
scientists, viii, 8, 9, 16, 31, 32, 34, 47, 48, 60, 71, 73, 77, 84, 88, 89, 92, 102, 113, 114, 118
sea level, 91
search, 47, 75
search engine, 75
searching, 44
Seattle, 6, 11, 29, 32, 34, 38, 50, 67, 77, 86
second party, 100
security, 104
seismic, viii, 2, 3, 4, 5, 8, 9, 10, 12, 13, 14, 15, 17, 19, 20, 22, 23, 25, 26, 27, 30, 31, 32, 33, 36, 37, 38, 44, 45, 49, 50, 53, 56, 57, 58, 60, 61, 64, 65, 66, 67, 71, 72, 73, 74, 76, 77, 78, 79, 80, 81, 84, 86, 87, 88, 89, 90, 91, 93, 94, 95, 97, 103, 105, 106, 115, 116, 118
seismic data, 14, 60, 89
Senate, 33, 36
sensors, 24, 27, 44, 57, 76, 84, 103
series, 9, 12, 55, 64, 65, 66
services, 52, 70, 104
severity, 4, 58, 62
shaping, 104
shares, 90
sharing, 81
shear, 45, 85, 99
Shell, 96
shelter, 21
shock, 27, 60
short-term, 15
signals, 60
simulation, 31, 37, 46, 72, 79, 88, 105
simulations, 23, 32, 38, 42, 51, 99
sites, 38, 58, 61, 77, 82, 85, 86, 88
social impacts, 21
software, 21, 22, 38, 45, 46, 52, 59, 70, 76, 78
soil, 31, 37, 45, 78, 80, 88
soils, 45
Solomon Islands, 82
solutions, 30

South Carolina, vii, 1, 3, 6, 65, 98, 110
spatial, 23, 60
spectrum, 31, 87
speed, 80
Sri Lanka, 92
St. Louis, 6, 11, 67, 70, 96
stability, 85, 88
Stafford Act, 71
stages, 39, 58, 75
stakeholder, 33
stakeholder groups, 33
stakeholders, 61, 73
standards, 19, 20, 33, 39, 45, 53, 58, 66, 73, 78, 87
Standards, v, 19, 32, 53, 64
State Department, 16, 51, 110
State Grants, 41
statutory, 31
steel, 37, 81, 88, 90
stock, 10, 26
strain, 44, 61, 83, 86
strategic, 39, 42, 98
strategic planning, 39
strategies, 54, 80
strength, 4, 45, 85, 115
stress, 22
students, 56, 71, 72, 80, 102
subscribers, 66
Sumatra, 48, 92
summer, 71, 77, 84, 85, 86
supply, 47, 103
supply chain, 47
surface wave, 45
survivability, 37
survival, 47
survivors, 88
sustainability, 92, 104
Switzerland, 60
systems, 9, 13, 26, 27, 37, 47, 56, 60, 77, 80, 89, 103

T

Taiwan, 8
talent, 79

targets, 73, 86
technical assistance, 93
technology, 43, 81, 87, 88, 104
technology transfer, 81
telecommunication, 17
telephone, 6, 16, 27
television, 59
television stations, 59
Tennessee, 6, 34, 63, 65, 67, 70, 79, 98, 99, 101, 110
territorial, 93
territory, 106
test data, 88
testimony, 64, 65
Texas, 28, 82, 86, 96
Thailand, 8, 92
threatened, 98
threatening, viii, 29
threats, 52
three-dimensional, 50
three-dimensional model, 50
threshold, 13
thresholds, 16
time, viii, 2, 5, 8, 13, 14, 15, 17, 22, 23, 33, 35, 57, 60, 61, 74, 75, 76, 77, 78, 82, 89, 90, 91, 92, 100, 113
timing, 15, 86
titanium, 57
tracking, 39
training, 32, 36, 38, 47, 52, 53, 61, 64, 65, 66, 67, 68, 78, 79, 90, 92, 93, 95, 98, 100, 105
transfer, 80, 83
transition, 76
transmission, 88, 98
transportation, 9, 26, 27, 45, 52, 61
tremor, 44, 71, 86, 88
tribes, 59
tsunami, 8, 27, 37, 59, 67, 79, 82, 88, 91, 92, 93, 97, 98, 100
Tsunami, 14, 73, 77, 91, 92, 93, 97, 100, 110
tsunamis, 5, 31, 51, 100

U

U.S. Agency for International Development, 92
U.S. Geological Survey, viii, 2, 3, 22, 31, 33, 34, 38, 83, 103, 110, 114
uncertainty, 12
undergraduate, 71, 80
uniform, 13, 14, 23, 47, 74
uninsured, 9, 10
United Arab Emirates, 82
United Nations, 92
United States, vii, viii, 1, 2, 3, 4, 5, 6, 7, 8, 9, 10, 11, 12, 13, 14, 15, 16, 17, 19, 22, 23, 26, 29, 30, 32, 38, 48, 49, 52, 61, 63, 66, 67, 69, 70, 73, 74, 78, 79, 86, 87, 89, 90, 91, 92, 96, 101, 108
universities, 15, 34, 44, 66, 79, 86, 97, 102, 105
updating, 32, 36, 37, 65
urban areas, viii, 4, 13, 29, 30, 74
urban centers, 73
urbanized, 14, 102
Utah, 6, 17, 60, 78, 86, 97, 99, 106, 110

V

Valdez, 91, 93
validity, 80, 99
values, 11, 85
variability, 5
vegetation, 43
velocity, 76
Vermont, 105
victims, 46, 64
Vietnam, 8, 92
Vietnamese, 68
virtual reality, 47
visualization, 47
voting, 104
vulnerability, viii, 2, 3, 47, 48, 94, 118

W

war, 89
War on Terror, 14
water, 56, 103
weakness, 26
wealth, 20
West Indies, 92
White House, 31, 35, 87
White House Office, 31, 35
wind, 87
winter, 12, 43

wireless, 79
Wisconsin, 4
women, 72
wood, 37, 81
workers, 60
workforce, 72, 81
working groups, 78, 99
Wyoming, 86, 106

Y

yield, 22